The Nature of the Atom

An Introduction to the
Structured Atom Model

The Nature of the Atom

An Introduction to the Structured Atom Model

J. E. Kaal

J. A. Sorensen

A. Otte

J. G. Emming

Curtis Press

©Stichting Structured Atom Model
Halvemaanweg 10, 7323 RW Apeldoorn, The Netherlands

ISBN (pbk) 978-1-8381280-2-9
ISBN (ebook) 978-1-8381280-4-3

Cover design: JudithSDesigns&Creativity, *www.judithsdesign.com*
Typeset by Falcon Oast Graphic Art Ltd, *www.falcon.uk.com*

Distributed in North America by SCB
Distributed in the United Kingdom and the rest of the world by
Gazelle Book Services Ltd

Visit Curtis Press at www.curtis-press.com

Contents

Preface

This book proposes a new model of the nucleus of the atom, a new model of the atom itself, and a way to bridge the current divide between chemistry and physics. The model is called the *Structured Atom Model*, in short SAM. In it the nucleus is static and has a defined structure.

SAM has been under development since 2006 by J. E. Kaal from the Netherlands. He was joined by J. A. Sorensen (Colorado, United States) and J. G. Emming (Utah, United States) in 2016. The latest addition to the team is A. Otte from Germany, who joined in 2019.

In the introduction we consider the rules of science we want to be measured against. There is nothing special about these rules, except that we do not follow the one rule that is never mentioned, never written down, never spoken about openly—the rule that currently dominates all other known rules of science, that is, work is scientific if and only if *it follows currently accepted theory*. If it does not do that and does not adhere to current wisdom, then this work has *by definition* severe scientific errors—it does not meet scientific standards and surely the proper scientific methods have not been used. What has been created is at best pseudoscience. This is the one, *most unscientific rule* of all, that sadly governs large parts of science today.

SAM is an empirical model of the nucleus of the atom founded on the intuitive belief that the atomic nucleus should have structural properties. Therefore, it will not be surprising that this book does not follow the pattern of most conventional physics books and that the rules of the Standard Model do not apply. To postulate that the nucleus has structure, automatically eliminates the idea that quantum mechanics is appropriate for the nucleus.

Not being constrained by currently accepted theory and shedding the burden of conventional thinking has the advantage that creativity can play a larger role than otherwise would be the case. On the other hand, we adopted from the outset the severe constraint that any geometrical structure we create must be in accordance with known material properties. This has compelled us to start with a clean slate and apply this new perspective to develop new rules for the growth of the elements.

Discovering and applying these rules has resulted in a model for the nucleus that is internally consistent and compatible with the known key properties of the elements, including whether they are stable or radioactive, their decay modes and rates, isotopic differences, etc. Given this new approach, the reader can expect to encounter several unconventional notions, some of which surprised even the authors. For additional visibility of a particular element or isotope we encourage the reader to use the Atom-Viewer (https://structuredatom.org/atomizer/atom-viewer) while reading the book. This tool gives a dynamic 3D view of the atomic nuclei.

July 2021   J. E. Kaal
                                                                                                   J. A. Sorensen
                                                                                                          A. Otte
                                                                                                   J. G. Emming

Acknowledgments

This book is the result of many years of work. The model and the book could never have materialized were it not for the many supporters across the world who have helped in their own personal way. Much support has been given by members of the Electric Universe (EU) community and also by EU organizers, giving us a slot for presentations during their conferences. Similarly, the Low Energy Nuclear Reactions (LENR) community has been kind enough to allow us to present the model during workshops.

One name in particular is important to point out. Norman Cook, who unfortunately passed away in 2019, has been the "torch carrier" of the idea that the nucleus has a specific organization which conflicts with the Copenhagen interpretation of quantum mechanics. His work through presentations, books, and involvement in LENR experiments is invaluable and deserves a place in the historical record. We hope that his legacy lives on as such and one day he will be recognized for this contribution to science.

We also want to thank our publisher, Neil Shuttlewood, for having the courage to take on such a project and for spending so much time making the book more readable.

In memory of Dr. Norman D. Cook (1949–2019)

Introduction

HISTORICAL CONTEXT

Since the discovery of the atomic nucleus science has been in search of its structure. Many nuclear models, based on quantum mechanics (QM), describe the nucleus in statistical terms thereby negating any fixed structure. The modified Rutherford/Bohr model consists of a nucleus made up of a featureless collection of protons and neutrons orbited by electrons. The protons and neutrons themselves are assumed to be moving in a gas-like manner, obeying the Heisenberg Uncertainty Principle, almost as if the nucleus were not a physical object. Although there is ample experimental evidence supporting the notion of a differentiated geometry of the nucleus, the Rutherford/Bohr model is still widely used for teaching purposes with QM being by far the dominant framework for analyzing the nucleus.

It is worth recalling the historical timeline of a specific circumstance related to the atomic nucleus. Prior to Chadwick's discovery of the neutron in 1932, the nucleus was thought to consist of a combination of protons and electrons. When the neutron was found to be distinct from the proton it was given independent status and the nucleus was then determined to consist of both protons and neutrons. Subsequent discussions of nuclear stability led to the introduction of the strong nuclear force, which heavily influenced the development of QM.

Given that we know the size of the nucleus for virtually all elements and isotopes—from ~1.7 fm (femtometers, 10^{-15} m) for the proton to ~15 fm for uranium—an obvious conflict arises with the notion of hundreds of nucleons moving freely in such restricted spaces. Therefore, in our view Heisenberg's Uncertainty Principle was incorrectly applied to the crowded atomic nucleus.

A NEW APPROACH

It appears that a new approach to this issue is required, one that provides clarity and avoids the need for multiple models—often operating under contradictory assumptions—to explain the various nuclear phenomena. To that effect, the *Structured Atom Model* (SAM) was created using two simple assumptions:

1 SAM doesn't distinguish between neutrons and protons. Thus, in SAM the nucleus consists of protons and electrons only. Essentially, this takes us back to 1932 before the neutron was introduced. We must be careful to distinguish these nuclear electrons, inside the nucleus ("inner electrons"), from electrons outside the nucleus.
2 The nucleus is kept together and given shape by the inner electrons positioned between protons, negating the need for the strong or weak nuclear forces. We call the resulting main principle "spherical dense packing." As the protons find their place naturally, based on the main principle, we would like to categorize this as an "unforced model."

Based on these two assumptions we can derive, in high detail, the shape of the nucleus of each element in the periodic table and their isotopes.

MOTIVATION

The authors of this book are engineers with a collective background in chemical engineering, electrical engineering, information technology, computer science, systems engineering, computer modeling, and space science instrumentation for ESA and NASA missions. As such we are outsiders and claim no specific expertise in atomic and nuclear physics, and this book about the structure of the atom therefore doesn't approach the subject from a conventional nuclear physics point of view. Rather, the underlying principles are more of an observational, geometrical, and logical nature than based on sophisticated mathematics. As outsiders in the field we have fewer inhibitions to challenge the status quo with a new paradigm than otherwise might be the case.

The SAM concept results from the conviction—based on many general observations—that the nucleus should have a recognizable structure with understandable properties. Since we know that, outside the nucleus, neutrons separate into protons and electrons, we arrived at the concept of a nucleus solely consisting of protons with electrons acting as "glue" between them, basically following the Coulomb force law. This lies at the heart of SAM, allowing the nucleus of each element to be dynamically visualized in a 3D tool, called Atom-Viewer.

OVERVIEW OF THE MOST IMPORTANT FINDINGS

The following list gives a broad overview of the most important findings that are carefully developed in this book:

- SAM is a tool that helps us consider nuclear structure. Specifically, we find that the properties of the elements can be directly tied to the structure of the nucleus. *In other words, the geometric structure of the nucleus and the physical and chemical properties of the elements are causally related.* In a certain sense this restores the relationship between chemistry and physics, after a divorce of more than a century.
- With the two rules for the nucleus, using only protons and electrons, and applying the principle of spherical dense packing, SAM has enabled us to reconstruct the elements virtually from scratch.
- We identified the growth pattern the nucleus adheres to when growing in size and number of nucleons. This pattern is fractal in nature and is made up of icosahedrons that are connected. The pattern shows a doubling of icosahedron structures on top of each new completed icosahedron.
- Branches can be detached from the nucleus in fission processes and become independent lighter elements themselves. In elements heavier than lead, branches can fuse together and break off as a result. This is known as the conventional fission process.

- Interference between branches of the nucleus can cause its structure to be stressed. This "stress" represents energy stored in the structure which can be accessed during fission.
- SAM allows us to identify specific locations for various nuclei where this stress originates. This structural phenomenon quite plausibly explains the origin of nuclear fission as well as the asymmetric breakup of the nucleus seen during fission processes.
- Closely related to this, we now recognize the fundamental cause of nuclear instability or radioactive decay and especially the importance of the role of the inner electrons in nuclear stability.
- Fission and fusion play a huge role in element creation, but not only in stars. Instead, we see those processes happen on planets as well as experimentally in laboratories under a specific set of circumstances.
- We pinpoint the structural explanation for the progression of the neutron/proton ratio in the periodic table. The elimination of the neutron has caused us to create a new numbering system for the periodic table based on the deuteron count.
- In the process of populating the new numbering system, we have flagged several possible structures that might be identified as "missing elements," since they do not fit in the standard periodic table.

To summarize, through SAM we:

1 establish a causal relationship between the nucleus and the "outer electron domain";
2 discover a source of potential energy stored in the nucleus of certain elements;
3 identify the structural cause for nuclear instability, nuclear fission, and radioactivity; and
4 identify several possible "missing elements."

REUNITING PHYSICS AND CHEMISTRY

Today, nuclear physics and chemistry are two distinct fields of science often considered only weakly connected.

Through its inclusion of astronomy, physics is the oldest academic discipline. Over much of the past two millennia, physics, chemistry, biology, and certain branches of mathematics formed part of natural philosophy. That changed during the scientific revolution in the 17th century, when disciplines became more or less independent.

Chemistry is the scientific study of matter composed of atoms, molecules, and ions. It deals with their structure, behavior, and the way in which they change during reactions with other compounds. It addresses topics like how atoms and molecules interact via chemical bonds to form new chemical compounds.

Chemistry used to be the dominant science up to the 18th century, in the 19th century physics grew stronger. The 20th century saw physics become the dominant science.

However, at what level does chemistry stop and nuclear physics start? Is there actually a boundary? Or is it all chemistry? From the definition of chemistry above, we understand that chemistry can deal with atomic structure as well. There is no a priori reason to exclude the nucleus and its structure. The creation of SAM is an attempt to bring nuclear physics and chemistry back together when contemplating the structure of the atom. Since in SAM the positioning of the protons as well as the inner electrons in the nucleus determines the positioning of the outer electrons, the "outer electron domain" is therefore causally connected to the inner structure of the nucleus. Thus, physics and chemistry are reconnected in SAM.

LOGIC AND THE SCIENTIFIC METHOD

Grammar, logic, and rhetoric were once the base of any classical education. The three subjects together were later denoted by the term "trivium." The tradition was already established in Ancient Greece. The term "quadrivium" denotes four subjects (arithmetic, geometry, music, and astronomy)—usually taught after the trivium.

This kind of classical education was lost in the last one and a half centuries and with it the sense of argumentation, thesis, theory, and what is right or wrong, at least formally that is. Does the setup of a theory adhere to the rules for proper science? Are the premises clearly spelled out? Do the conclusions logically derive from the premises?

Or, to phrase it differently: What is a good scientific result? What are good methods? Such questions have at least logical, empirical, and historical answers. Foremost the results of such a scientific methodology must be:

- *consistent* (both internally and externally);
- *parsimonious* (sparing in proposed entities, explanations; commonly known as Occam's razor);
- empirically *testable* and *falsifiable*;
- based upon *controlled* and *repeated* experiments;
- *correctable* and *dynamic* (changes are made to it when new data becomes available);

but also (although to a much lesser degree):

- *useful* (able to describe and explain *observed* phenomena);
- *tentative* (not asserting absolute certainty);
- *progressive* (achieving all that previous theories have, and more).

This is what theories need to be measured against—not only new theories but also those already established. The loss of the classical approach to education has meant that there are many accepted theories that violate at least one, in most cases more than one, of the rules of good science. When we follow those basic rules, we can find truthful answers. Other methodologies are questionable. There are no shortcuts.

The classical "trivium" approach (grammar, logic, and rhetoric) is a sensible approach to scientific work, that is, first we identify facts, then definitions, and finally make observations. Next, we can apply logic to our observations by taking more than one observation into account at a time. After making enough logical connections we can attempt to express this all through rhetoric—this book in the case of SAM. To deny an observation, simply because it does not fit within our established model, is not an option. In that case we have to change the model.

Why do we think it is important to talk about logic and scientific methodology in the introduction of this book? Because something important got lost in the last few decades and "nonscience returned" as a result of this loss. We once understood the importance of proper science, and we need to get that understanding back.

THIS IS A WORK IN PROGRESS

We also need to recognize that we simply do not have all the answers when trying to explain the atom with this new model. Too much of what we think we know is based on assumptions—whether right or wrong (i.e., our current models and theories may be fundamentally flawed). We need to recognize this and be humble—accepting that there may be errors in our discoveries. The authors of this book are convinced a new and improved understanding of the fundamentals is required before we can even attempt to understand the world around us.

This book is a work in progress. Please take it as such. There is a lot we do not yet know but think that there is enough material about SAM available to present what we have. Much more research needs to be done, a long list of which appears in Appendix E. If you—the reader—think there is something in this then join us in our research.

HOW TO READ THIS BOOK

After a few introductory chapters in Part I, starting with Chapter 2, we delve deep into the details of the model. Chapters 2 to 5 have summary sections at the end of each chapter. If the content of the chapters becomes too detailed, we recommend reading at least the summary sections at the end of those chapters as well as the whole Chapter 6, as it is an interim summary of Part I.

The information contained in Part II is a kind of payoff to all the work done in Part I, as we further underpin the model, compare it with other models of the nucleus and apply it to real world issues. If there are questions about some of the details of the model, return to those details in Part I after you have read Part II. Make use of the glossary in Appendix K of the book if the terminology is unclear.

The themes of this book are sometimes developed and elaborated in a spiral approach—they might be considered several times throughout the book from different vantage points or (more likely) when connections are made to other subjects.

We use *Wikipedia* extensively as a source for definitions and generic Standard Model explanations. This is on purpose to allow our readers easy access to background information related to the topics addressed in this book.

PART I

The Master said, "You, shall I teach you what knowledge is? When you know a thing, to hold that you know it; and when you do not know a thing, to allow that you do not know it—this is knowledge."

Confucius, Analects

CHAPTER 1

Setup

1.1 THE RUTHERFORD/BOHR MODEL OF THE ATOM

Most people will remember the Bohr model of the atom from their high school days—a system with a small, dense nucleus surrounded by orbiting electrons. The nucleus consists of positively charged objects—later differentiated into protons (positive) and neutrons (neutral)—in an unstructured blob with negatively charged electrons in the outer regions of the atom—orbiting like planets around the Sun. The model was envisioned in 1913 by Niels Bohr as an extension of Rutherford's model (1911). The Rutherford/Bohr model (Fig. 1.1) is a relatively primitive model, compared with the valence shell atom model that was later developed (or derived) from it.

But this very simple model is still being used in education. Of course, we are told that it doesn't represent reality. However, not many people realize how many models of the atom have been created that continue to be used to this day. We will look at the basic assumptions of some of them in Section 7.1, for now it is enough to remember the basic structure of the Rutherford/Bohr model.

In the context of the Rutherford/Bohr model and others that followed it, an *element* is defined by the number of protons in the nucleus and an equal number of electrons "orbiting" the nucleus, making it neutral overall.

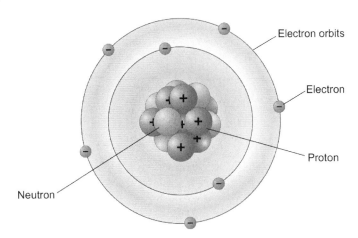

Figure 1.1 Bohr atomic model.
[Encyclopedia Britannica 2021/science/Bohr-model]

1.2 THE NEUTRON

When James Chadwick *discovered* the "neutron" in 1932 there was great debate about whether the nucleus was composed of protons and "nuclear electrons" or protons and "neutrons." At the *7th Solvay Conference* in Brussels, Belgium, in 1933, discussion resulted in the neutron being assigned the status of a fundamental particle. The main issue was that nobody at that time could solve the mystery of how "nuclear electrons" could be trapped in the nucleus. Instead, it was decided that the neutron was not a combination of a proton and an electron (this is however what "it" decays to after 15 minutes outside the nucleus) but rather a fundamental particle without an electrostatic charge, meaning it is neutral. The decision resulted in the invention of the strong nuclear force, needed to hold protons in the nucleus because protons repel one another.

In SAM, we don't subscribe to the 1933 Solvay decision, instead we postulate that the neutron is not a fundamental particle in the nucleus at the same level as the proton. *Thus, SAM views the nucleus as comprised of protons and nuclear electrons. These electrons keep the protons together in a fixed geometric structure—based on the principle of spherical dense packing*—which we will discuss in Section 1.4. In this sense the nuclear electrons are a kind of "glue" holding the protons together in a structure. The result is a nucleus with an overall positive charge. We put aside any possible issues raised by this concept of a proton–electron nucleus, postponing discussion to Sections 8.2 and 8.3.

One other thing is already a clear consequence of this concept: we will have to re-define the term "element." We can no longer use the number of protons to distinguish between elements. It remains to be seen if we can keep the number of outer electrons as a defining criterion for elements.

1.3 INITIAL DEFINITIONS

Before proceeding, we must clarify the most basic terms and concepts used in this book:

- Whenever the term "proton" is used, its definition in the model is as a positively charged particle, most likely taking a spherical form. All protons have the same properties and cannot occupy the same space.
- Whenever the term "electron" is used, its definition is a negatively charged "something" forming a duality with the proton. There are at least two "states" an electron can assume: an "inner electron" state and an "outer electron" state. An outer electron is what is currently known as an "orbital" electron—associated with chemistry. The shape of an inner electron (nuclear electron) is not yet known, it might depend on its position in the nucleus. Also, we must consider the possibility that there could be several types of "inner electrons" based on their position in the nucleus. An electron has mass, but we cannot be sure it is matter.
- A "proton–electron pair" (PEP) is a proton with a closely tied inner electron. It is not a fundamental particle at the same level as the proton. A PEP is also *not* hydrogen-1 where the electron is an outer electron. The term "proton–electron pair" is our own

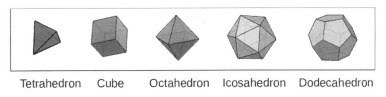

Tetrahedron   Cube   Octahedron   Icosahedron   Dodecahedron

Figure 1.2 The five platonic solids.

creation. Whenever the term "neutron" is used, the reader should realize that it (the so-called neutron) is actually a proton–electron pair, not a fundamental particle.

- A "nucleon" is a particle in the nucleus. In SAM a nucleon is synonymous with the proton since we have eliminated "neutrons" as separate entities. We do not currently count the inner electrons as nucleons, because we do not think an inner electron qualifies as a particle.
- A "deuteron" is defined as two protons combined with one "inner electron" positioned between them. This is the essential building block recognized in SAM. It is obviously positively charged.
- A "single proton" is a nuclear proton that is neither part of a PEP nor part of a deuteron.
- An "element" is characterized by the shape of its nucleus for a given number of deuterons and single protons (protons not bound as part of a deuteron) it contains. The same number of outer electrons makes the atom a neutral object.
- The "element number" of an element is the number of deuterons plus the number of single protons in its nucleus. The element number does not hold any information about the shape of the nucleus, it most likely will not uniquely define an "element."
- An "isotope" is a variation of an "element," in most cases by the addition of extra PEPs to the nucleus. This leaves the element number unchanged. In some cases an element can also lose a PEP to create an isotope.
- A "nuclear isomer" is a variation of an isotope having the same components (same number of deuterons plus single protons, same number of PEPs) but arranged in a different way and/or with different energy levels.
- A "platonic solid" is constructed from polygonal faces, identical in shape and size (congruent) as well as having equal sides and angles (regular) and with the same number of faces at each vertex. Five solids meet these criteria: the tetrahedron (four vertices, four faces); the cube (eight vertices, six faces); the octahedron (six vertices, eight faces); the icosahedron (twelve vertices, twenty faces); and the dodecahedron (twenty vertices, twelve faces) (Fig. 1.2).

Are protons and electrons composed of something smaller? For our model this is not relevant. The model starts with the proton and electron recognized as being fundamental.

Additional definitions will be introduced in the book when needed.

1.4 STRUCTURES WITH SPHERICAL DENSE PACKING

In SAM, the nucleus is constructed by putting protons (positive charge) together in a geometrical structure with electrons (negative charge) inside this structure at very specific positions (and yet with unknown form and size), acting as a sort of "glue." The development of the structure then follows the rule of "spherical dense packing," which manifests like an attracting force to an imaginary center of the structure. It seems likely that platonic solids play a role in dense packing. This only works for those three platonic solids that have triangular faces—the tetrahedron, octahedron, and icosahedron. The central structure of the octahedron is a square with spheres stacked on top of one another. The octahedron is therefore not stable as the stacking allows for easy movement. The icosahedron and tetrahedron are the only two platonic solids that are stable and not easily broken. There exists at least one densely packed structure without being a platonic solid. We will encounter it when looking at seven spheres.

The following observations can be made while building up "spherically, densely packed" structures one sphere after another:

1. A single sphere exists by itself. We could say that it has zero dimensions.
2. Two spheres can attach to each other in only one way (i.e., next to each other). This gives direction and depicts one dimension.
3. With three spheres there are two possibilities—in a line which represents one dimension or in a triangular shape in two dimensions. Three spheres in a line is not, of course, a densely packed structure.
4. Four spheres have three possibilities—all four in a line (one dimension), in the shape of a square (two dimensions), and in the shape of the tetrahedron (three dimensions; Fig. 1.3). Four in a line doesn't work because it is not a spherically dense structure, neither is the square. This leaves the tetrahedron, the first shape that is three-dimensional in our lineup. Such a structure is densely packed and quite stable as it cannot be easily crushed or distorted. It is also the simplest of the platonic solids.
5. A fifth sphere cannot be added without distorting the structure of the tetrahedron, with the inward-pulling force breaking it. There is no stable structure with five spheres. The spheres push each other out of a dense structure and create distortions in the structural shape—there is no final resting shape for clusters of five spheres.
6. Six spheres create an octahedron, and at first glance this appears to be stable and densely packed. However, since the spheres are being pulled toward a common center, opposite spheres will push apart the four connected protons and instead create a five-sphere ring structure with an additional sphere on one side at the center of the ring. This structure is therefore not balanced.

Figure 1.3 Four spheres in dense packing—a tetrahedron.

7 Seven spheres come together to form a *pentagonal bi-pyramid* consisting of a ring of five spheres and a top and bottom sphere to cap the ring (Fig. 1.4). This shape has 10 triangular faces and many of the attributes of a platonic solid. The *pentagonal bi-pyramid* can be thought of as a partial *icosahedron*. We can build an icosahedron with spheres by creating two rings of five, putting them together into a cylinder of two interlocking rings, then adding one more sphere to each end. The pentagonal bi-pyramid is the same shape as the icosahedron minus one of the two rings of five.

Figure 1.4 Seven spheres in dense packing—a pentagonal bi-pyramid.

8 Eight spheres create the same conditions found with five spheres. The inward-pulling force causes the 8th sphere to penetrate the ring below it, breaking the underlying structure. It is not possible to create a stable structure that is densely packed with eight spheres.

9 When adding two spheres on top of the pentagonal bi-pyramid (seven spheres) we create a new tetrahedron that combines with two spheres already in the structure. This tetrahedron is stronger than the underlying structure and modifies it accordingly; therefore, the underlying structure is not destroyed as happened when adding the 8th sphere.

10 The 10th sphere creates the next viable structure. However the result is not symmetrical, yet stable.

11 The 11th sphere completes the next tetrahedron structure, but it is still not entirely symmetrical.

12 The 12th sphere creates a structure with an icosahedron shape—the largest platonic solid that can be densely packed (Fig. 1.5). The icosahedron has 20 faces, each forming an equilateral triangle with 3 protons at the vertices. This is a very stable structure. Even though it follows the rules for spherical dense packing toward a common center it is hollow on the inside.

The creation of even bigger conglomerations follows the rules outlined above and repeats the basic tetrahedron, pentagonal

Figure 1.5 Twelve spheres in dense packing—an icosahedron.

bi-pyramid, and icosahedron structures with various intermediary steps. Those first 12 configurations are therefore the basic structures forming bigger conglomerations. Their stability will be considered in more detail in Section 2.2.

1.5 INTRODUCING SOME SIMPLE NUCLEAR REACTIONS

Some nuclear reactions are very well known and have been thoroughly studied. One would expect that we therefore know all there is to know about such reactions and that we understand them properly, after all they are being used routinely in nuclear power plants and nuclear warhead tests.

In the language of the Standard Model, when a neutron in the nucleus is converted into a proton a high-speed electron is emitted as well as an anti-neutrino. This is called β– decay (beta minus). If a proton in the nucleus is converted into a neutron then a high-speed positron (an electron with a positive charge) is emitted as well as a neutrino (Fig. 1.6). This is called β+ decay (beta plus).

In SAM, there is no fundamental neutron. Therefore, if a proton–electron pair (PEP) in the nucleus loses an electron the proton remains while the electron is ejected from the nucleus (β– decay). This can occur if two PEPs combine to create a deuteron (two protons with one electron in between). This can also happen when there is no room in the nucleus to hold all inner electrons—when there is too much negative charge in the nucleus. As a consequence, one electron is emitted and a deuteron is created.

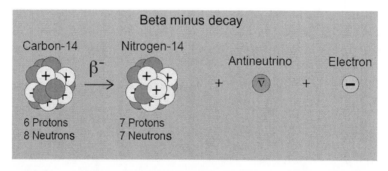

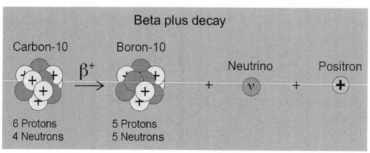

Figure 1.6 Example β– and β+ decay schemes.

[https://www.shmoop.com/study-guides/physics/modern-physics/particle-physics]

The superfluous electron moves out of the nucleus and most likely out of the atom. Perhaps it will be captured in the area of the outer electrons, dependent on its kinetic energy.

Let us now consider β+ decay. For example, when a deuterium atom (a deuteron nucleus plus one outer electron) is being broken up, an outer electron moves into the nucleus and two separate PEPs are created. Another example would be where there is just too much positive charge in the nucleus, providing the opportunity for an outer electron to become an inner electron. In SAM the involvement of a so-called positron in this decay scheme is not needed. To some this process might look like a positron is being emitted, but in SAM it is always an outer electron being integrated into the nucleus. In SAM there is no real distinction between β+ decay and electron capture, it is always an incoming electron maybe combined with resettling of the nucleus. We will look at these questions in more detail in Section 11.4. Considering SAM, it is also questionable whether neutrinos or anti-neutrinos are involved in these processes—the neutrino was initially introduced here by Wolfgang Pauli to "fix" a calculation of spin and its introduction can therefore be seen as artificial, perhaps lacking physical basis. We will discuss spin, what it means in SAM, and how it is determined in Section 5.3. Also, the continuous β– decay energy spectrum, another subject of Pauli's "fix," will be considered in Section 11.2.2.

An α (alpha) particle is the nucleus of a helium atom. We will see that it is shaped as a tetrahedron according to the rules of spherical dense packing. When such a particle is emitted from a nucleus, it is called α decay.

There is also γ (gamma) decay. This type of decay results in the emission of a photon (in the electromagnetic spectrum) from the nucleus. It is also possible for photons to interact with protons in the nucleus, but usually they interact with outer electrons. The latter however is not a nuclear reaction, as the nucleus is not structurally changed.

Together the processes described above are called radioactive decay. A material containing unstable nuclei is understood to be radioactive and radioactive decay is considered a random process at the level of single atoms. However, is it really? We will come back to this question in Section 11.1.

Overall, we now have the following simple nuclear reactions available to us:

α decay: emitting a helium nucleus from the nucleus
β+ decay: adding an electron to the nucleus (e-capture)
β– decay: emitting an electron from the nucleus
  +PEP: absorbing a PEP into the nucleus (also called n-capture/neutron-activation)
  –PEP: emitting a PEP from the nucleus.

The last two reactions in this list are mentioned in the process of creating isotopes. The –PEP is of interest, as it corresponds to the emission of a "neutron" from the nucleus. Outside the nucleus, the "neutron" itself is unstable (with a half-life of roughly 15 minutes) and decays with a β– step into a proton and an electron.

Furthermore, we will assume that as there is electron emission and electron absorption and also PEP emission and PEP absorption, there should also exist in principle proton emission from the nucleus and proton absorption (proton capture) into the nucleus. We assume this despite the existence of the so-called Coulomb barrier of the nucleus, which is electrostatic in nature, assumed to prevent this kind of interaction. We will consider the topic of proton capture (p-capture) in Section 11.11.

Of course, we must also—for completeness—ask whether there is some kind of absorption/fusion of α particles into the nucleus (for a more in-depth consideration of simple nuclear reactions see Chapter 11 in Part II).

## CHAPTER 2

Developing the elements in SAM

The periodic table of elements (PTE) is an attempt to organize and group the known elements into a logical system. Original versions showed vacant spots and therefore unknown elements. In 1789 Antoine Lavoisier started with a list of 33 elements which he grouped into gases, metals, non-metals, and earths. The next 80 years were spent attempting different organizational schemes. Johann Wolfgang Döbereiner, Leopold Gmelin, Jean-Baptiste Dumas, August Kekulé, Alexandre-Émile Béguyer de Chancourtois, Julius Lothar Meyer, William Odling, John Newlands, and Gustavus Hinrichs all worked on this task of finding different organizational schemes. However, Dmitri Mendeleev's publication in 1869 is considered the first usable PTE (Fig. 2.1). In his table he left gaps for yet undiscovered elements, predicting their properties according to their positions. He also ignored atomic weight discrepancies in favor of classification into chemical families. The typical representation we currently use was created by Horace Groves Deming in 1923.

We will now start to create the nuclear structure of the first elements in the PTE according to the rules stated before (spherical dense packing, protons and inner electrons, and no neutrons). Here we will not initially consider the positioning of the inner or outer electrons. As stated earlier, an element is defined—according to SAM—by the shape of its nucleus for a given number of deuterons and single protons not bound in deuterons and an equal number of outer electrons.

A nucleus can grow by adding a proton, a proton–electron pair (PEP), a whole deuteron, or several of these. Additionally, a PEP might enter the nucleus, break up, or decay, and provide only a proton to the nucleus, with the electron being ejected and sometimes becoming an outer electron. The number of deuterons and single protons plays a special role in the buildup of the elements as they define the element itself in combination with the shape of the nucleus.

In the buildup of atomic nuclei the question is always, where will the next proton, PEP, or deuteron be placed on the nucleus?

One rule we have already noticed while looking at spherical dense packing in general is that buildup is about creating a tetrahedral structure every time we add another sphere. If this tetrahedral sub-structure cannot be created through the addition of a sphere, then the new sphere will not become part of the existing structure, it will destroy the structure instead (Section 1.4).

— 70 —

но въ ней, мнѣ кажется, уже ясно выражается примѣнимость вы ставляемаго мною начала ко всей совокупности элементовъ, пай которыхъ извѣстенъ съ достовѣрностію. На этотъ разъ я и желалъ преимущественно найдти общую систему элементовъ. Вотъ этотъ опытъ:

			Ti=50	Zr=90	?=180.
			V=51	Nb=94	Ta=182.
			Cr=52	Mo=96	W=186.
			Mn=55	Rh=104,4	Pt=197,1
			Fe=56	Ru=104,4	Ir=198.
			Ni=Co=59	Pl=106,6	Os=199.
H=1			Cu=63,4	Ag=108	Hg=200.
	Be=9,4	Mg=24	Zn=65,2	Cd=112	
	B=11	Al=27,4	?=68	Ur=116	Au=197?
	C=12	Si=28	?=70	Sn=118	
	N=14	P=31	As=75	Sb=122	Bi=210
	O=16	S=32	Se=79,4	Te=128?	
	F=19	Cl=35,5	Br=80	I=127	
Li=7	Na=23	K=39	Rb=85,4	Cs=133	Tl=204
		Ca=40	Sr=87,6	Ba=137	Pb=207.
		?=45	Ce=92		
		?Er=56	La=94		
		?Yt=60	Di=95		
		?In=75,6	Th=118?		

а потому приходится въ разныхъ рядахъ имѣть различное измѣненіе разностей, чего нѣтъ въ главныхъ числахъ предлагаемой таблицы. Или же придется предпо- лагать при составленіи системы очень много недостающихъ членовъ. То и другое мало выгодно. Мнѣ кажется притомъ, наиболѣе естественнымъ составить кубическую систему (предлагаемая есть плоскостная), но и попытки для ея образо- ванія не повели къ надлежащимъ результатамъ. Слѣдующія двѣ попытки могутъ по- казать то разнообразіе сопоставленій, какое возможно при допущеніи основнаго начала, высказаннаго въ этой статьѣ.

Li	Na	K	Cu	Rb	Ag	Cs	—	Tl
7	23	39	63,4	85,4	108	133		204
Be	Mg	Ca	Zn	Sr	Cd	Ba	—	Pb
B	Al	—	—	—	Ur	—	—	Bi?
C	Si	Ti	—	Zr	Sn	—	—	—
N	P	V	As	Nb	Sb	—	Ta	—
O	S	—	Se	—	Te	—	W	—
F	Cl	—	Br	—	J	—	—	—
19	35,5	58	80	190	127	160	190	220.

Figure 2.1 The periodic table from Dmitri Mendeleev's *Osnovy khimii* (1869; *The Principles of Chemistry*).

Does nature develop all elements and isotopes by adding small chunks (PEPs, deuterons)? We simply don't know the answer. However, adding one or more PEPs ("neutron" capture) and then getting rid of an electron by β– decay are both known, simple nuclear reactions, widely observed in nature. We will follow this path and see where it takes us.

Obviously, it is not only the number of protons, but also the number of deuterons, in the nucleus that is relevant to the buildup in SAM, and while the number of protons might be the same (what we considered when talking about spherically dense packing in general) the number of inner electrons might differ in some cases and therefore change the stability of a nucleus. To be clear on this, often an element is not only known by its name, but also by—as defined in SAM—the number of protons (traditionally protons and "neutrons") in its nucleus (i.e., gold-197, carbon-12, or iron-56). In SAM the trailing number represents the total number of protons in a nucleus.

2.1 LOOKING AT THE FIRST ELEMENTS

We start with hydrogen-1, which in itself is odd. It is essentially an isotope of deuterium (hydrogen-2), but the number of deuterons in the nucleus is zero. It is just one proton in the nucleus and one outer electron. Hydrogen-1 can be considered as some kind of proto-matter. It represents the case of one sphere in our buildup of densely packed spheres (Section 1.4).

In order to move to deuterium (two spheres), we have to add a PEP to the nucleus. Now we have two protons connected by an electron between them, a *deuteron*. Still, there is only one outer electron. A proton on its own could not be added, the "glue" between them would be missing, and they would not stay together. Adding only an electron would make the nucleus neutral. This is not an option because nuclei need to have a positive charge, to be offset by the outer electrons with negative charge.

> The PEP is the only option here; negatively charged nuclei do not exist, nor do neutral nuclei—with the exception of the PEP itself, which is not a nucleus.

According to the rules of spherical dense packing there are two options for three spheres: three in a row or positioned in a triangle. This corresponds to the options we have in this case: we can either add a PEP to the nucleus or just a proton. There is definitely no room to add another electron. Adding a PEP would keep the nucleus positive, by one, balanced by the one outer electron. Hydrogen-3 or tritium is represented by the three-in-a-row case, the only option to have two electrons in the nucleus with three protons. When discussing the dense packing we stated that this line configuration—although possible—is not optimal. Indeed, tritium is not stable and only exists in trace amounts (Fig. 2.2).

The triangular case (adding only a proton) would only allow room for one (possibly central) inner electron, but we would also need a second outer electron to make

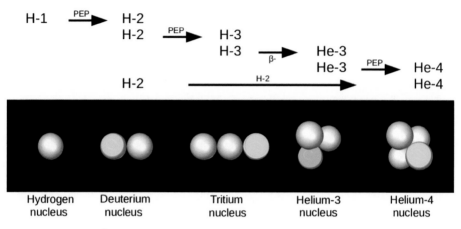

Figure 2.2 Nucleus buildup from hydrogen to helium.

this construct neutral. This is, of course, helium-3, an isotope of helium with one PEP missing from the nucleus. We could arrive at helium-3 by β– decay from hydrogen-3.

Helium-4 consists of two deuterons—each one being two protons glued together by a middle inner electron—aligned in a tetrahedron shape. This leaves another possible option for arriving at helium, that is deuteron–deuteron fusion (deuterium fusion). In Figure 2.2 we show the buildup through PEPs and β– decay. The light-yellow marked spheres show an added PEP, light green signifies rearranged spheres.

We have now reached the first platonic solid—the tetrahedron shape. The next sphere to be added will destroy this structure—as established earlier when discussing spherical dense packing. Indeed, two expected isotopes with five protons, named helium-5 and lithium-5, have half-lives of 700×10^{-24} s and 370×10^{-24} s, respectively. They are artificial and decay immediately if created by force.

Let's assume for now that we are able to simultaneously add two spheres to helium-4, with for example a deuterium atom or a deuteron (nucleus of deuterium), again deuterium fusion. Now we have three deuterons in the nucleus and the outer ones can share the middle one to come close to two tetrahedron shapes (Fig. 2.3).

Lithium-6 is a stable configuration, but there is a price to pay: a gap opens on one side of the nucleus which means it is not an optimal configuration and is not preferred—the gap asking for the addition of another sphere to close it. This can be achieved by adding another PEP to the lithium-6 structure to arrive at lithium-7, thereby morphing the structure into a very stable pentagonal bi-pyramid shape. This is not a platonic solid, but it is close. It can be described as a ring of five spheres with top and bottom capping spheres. There is a gap between the top and the bottom spheres, but it is on the inside of the structure and is thus shielded. Interestingly, both lithium-6 and lithium-7 are stable.

As stated earlier (Section 1.4) the six spheres structure (lithium-6) has a gap, which is apparently closed most of the time—according to the abundance numbers of lithium isotopes. Lithium-7 is much more abundant than lithium-6.

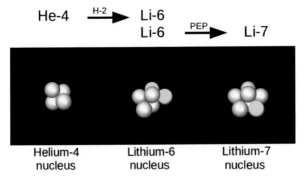

Figure 2.3 Nucleus buildup from helium to lithium.

Adding a PEP changes the ratio of inner electrons to protons in the nucleus. With the lighter elements we initially see a charge ratio of 1 inner electron (negative charge) to 2 protons (positive charge). If we add proton–electron pairs (PEPs) to this structure, we add inner electrons and protons in the charge ratio of 1 to 1, therefore also bringing the charge ratio of the whole structure closer to 1 to 1. Adding PEPs, we therefore are beginning to see a buildup of negative charge in relation to the initial ratio.

Adding another PEP brings us to eight spheres—a situation similar to five spheres. The pentagonal bi-pyramid shape of lithium-7 is distorted in a way that does not allow for a stable configuration. Lithium-8 decays first to beryllium-8 by electron ejection (β– decay). Although this ejection relieves some stress associated with negative over-charge in the nucleus—as with lithium-8—it is still unstable. Beryllium-8 further decays by α decay into two helium-4 atoms, having both stability and being a preferred configuration (Fig. 2.4). This is a dead end.

Again, we have to assume that it is possible to add two PEPs to the nucleus of lithium-7 at once to end up with lithium-9, or we have to add a whole deuteron. Nine spheres are in theory a stable configuration, but lithium-9 shows too big a negative charge buildup in the nucleus, thereby causing β– decay and eliciting a structural rear-rangement into beryllium-9—a stable structure comprised of tetrahedrons with shared protons (Fig. 2.5).

What is the difference between lithium-8/9 and beryllium-8/9? Structurally beryl-lium looks denser than lithium. The reason is the number of deuterons in the nucleus: lithium has three, beryllium has four. There is one more inner electron in beryllium that acts as "glue."

You can add one more PEP to beryllium-9 to arrive at beryllium-10. Ten spheres are a stable, but asymmetric configuration, according to the rules of spherical dense packing. Also, we see again too big a negative charge buildup in the nucleus—the re-sulting β– decay relieves stress and also rearranges the nucleus into boron-10 (Fig. 2.6).

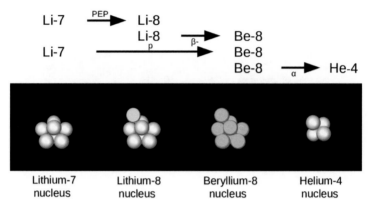

Figure 2.4 Nucleus buildup from lithium to beryllium (first attempt).

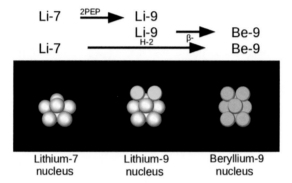

Figure 2.5 Nucleus buildup from lithium to beryllium (second attempt).

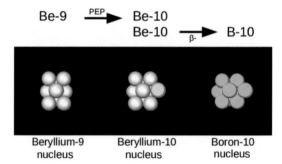

Figure 2.6 Nucleus buildup from beryllium to boron.

Adding another PEP takes us to boron-11, a stable, but still asymmetric, configuration that is also more abundant than boron-10. Addition of a further PEP takes us to boron-12. Again we see too big a negative charge buildup to keep the nucleus stable, the resulting β– decay relieves stress and also rearranges the nucleus into carbon-12—a

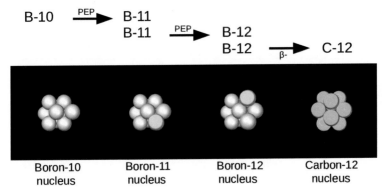

Figure 2.7 Nucleus buildup from boron to carbon.

platonic solid, the icosahedron, a very much preferred dense packing configuration (Fig. 2.7). There is also another outcome of boron-12 decay. In 1.6% of cases, it reacts with an α decay to beryllium-8, which then decays further into two more α particles. This hints at two different versions of boron-11/12 (different PEP placement), where one can create the next deuteron and the other one cannot.

By now a pattern of growth has become clear. However, we have reached the end of the pattern that was described by spherical dense packing (Section 1.4). Before we consider the next steps let's first reflect on what we have learned so far.

2.2 ELEMENT STABILITY AND ABUNDANCE

Some elements and isotopes are stable, others are unstable. Some are abundant, some are rare or artificial. When we discussed the ramifications of spherical dense packing (Section 1.4) we noted that not every configuration of spheres is structurally sound and those resulting from dense packing are sometimes not symmetric, or not balanced. Is element stability and abundance linked to structure? It appears so. Let's give this some consideration and make a side-by-side comparison (Table 2.1) of what we learned about dense packing and the first isotopes and elements.

The tetrahedron structure and helium-4 are both stable, having a preferred configuration. We also conclude that the dense packing would not allow for isotopes with five protons, and indeed, nothing stable exists with five spheres. The same is true for eight spheres. The structural deficits of six spheres versus seven spheres we can find reflected in their abundance. And those are just some examples. There appears to be a correlation between the results of spherical dense packing and known element attributes, at least for this first batch of elements up to carbon.

Table 2.1 Stability comparison between the packing scheme and the first isotopes.

Number of spheres	Isotopes and elements	Comments on spherical dense packing	Comments on abundance and stability of elements and isotopes
1	Hydrogen-1	Just one sphere, stable	Stable, highly abundant
2	Hydrogen-2	Line of two, stable	Stable, rare
3	Hydrogen-3	Line of three spheres, not very optimal, not densely packed	Unstable, only trace amounts
	Helium-3	Triangle, not optimal, stable	Stable, very rare
4	Helium-4	Tetrahedron structure, platonic solid, very stable	Stable, close to 100% abundant
5	Helium-5	Structurally unsound	Very unstable, artificial
	Lithium-5	Structurally unsound	Very unstable, artificial
6	Lithium-6	Structure has a gap ready to receive another sphere, otherwise structurally sound	Stable, but not abundant
7	Lithium-7	The gap is closed, very stable structure (pentagonal bi-pyramid), close to a platonic solid	Stable, abundant
8	Lithium-8	N/A	Very unstable, artificial
	Beryllium-8	Structurally unsound	Very unstable, artificial
9	Lithium-9	N/A	Very unstable, artificial
	Beryllium-9	Stable structure, created out of tetrahedrons with shared spheres	Stable, 100% abundant
10	Beryllium-10	N/A	Close to stable, only trace amounts
	Boron-10	Structurally sound, but not symmetrical	Stable, but not very abundant
11	Boron-11	Structurally sound, but not symmetrical	Stable, abundant
12	Boron-12	N/A	Very unstable, artificial
	Carbon-12	Icosahedron structure, platonic solid, very stable	Stable, abundant

2.3 PREFERRED CONFIGURATIONS

Some configurations of isotopes are preferred over others. If there are structural options, the abundance of isotopes of elements is not equally distributed, some are clearly more abundant than others. But which and why? To give at least part of the answer from observations, we create a minimal sphere around the nucleus which encompasses all the protons, a so-called minimal bounding sphere (MBS). Let us assume that our spheres have a radius of 1, then we get the following values (Table 2.2) for the minimal bounding spheres (radius and center computed using the CGAL 5.0.2 library [Fischer et al. 2020]), based on the SAM nucleus. There is also one other topic to consider, the so-called *binding energy* (BE) of the nucleus. This is the energy required to break up the nucleus into its fundamental components. Therefore, the binding energy represents the energy released when the nucleus came together, creating deuterons and the specific structure. At this point, how such values are calculated or measured is unimportant and is a topic for Section 7.3—for now we will just take them at face value. When comparing the BE of two elements in a transmutation step, or as is shown in Table 2.2 with a progressive buildup of the elements, a higher BE for the resulting element means that energy is released in the process. The other values are sampled from official *International Atomic Energy Agency* [IAEA 2021] data. The "PEP gap" marking in the "Additional PEPs" column denotes the portion of PEPs that fill a preferred spot (commonly known as a neutron gap).

The data in Table 2.2 shows that a smaller sphere (MBS radius) is not always preferred (e.g., step from beryllium-10 to boron-10). Also, the number of protons in the nucleus appears to play a role, a kind of volume per proton ratio of the bounding sphere (MBS vol./#p). What does this mean? It is probably an energetically preferred state. However, there is something else we mentioned earlier, that is, adding PEPs to a nucleus changes the ratio of inner electrons (negative charge) to protons (positive charge). This negative overcharge in relation to the normal 1 to 2 ratio in the nucleus, is relieved by electron expulsion (β– decay). Two additional PEPs appear to be enough to trigger this mechanism for the first elements unless the structure provides a gap ready to receive a PEP. Lithium-6 is an example with a gap, consequently lithium-7 can absorb two PEPs to arrive at lithium-9 before decaying. Of all the isotopes of lithium, lithium-7 has the best (lowest) MBS volume per proton ratio.

It is clear that helium-3 is energetically preferable to hydrogen-3, as it has a much better (lower) volume per proton ratio (it is more densely packed!), but there seems to be no urgency to this process—hydrogen-3 has a half-life of 12.32 years. To explain this we need to remember that tritium is made up of two interconnecting deuterons, which are stable connections. Additionally, the BE goes down from hydrogen-3 to helium-3. This might also prevent the move to helium-3 as energy needs to be spent.

Table 2.2 Some selected nuclear data for the first isotopes.

Isotope	Additional PEPs	Half-life	MBS radius	MBS vol./#p	BE (MeV)
Hydrogen-1	−1	Stable	1.00000	4.1888	N/A
Hydrogen-2	0	Stable	2.00000	16.7552	2.2250
Hydrogen-3	+1	12.32 y	3.00000	37.6991	8.4820
Helium-3	−1	Stable	2.15023	13.8810	7.7180
Helium-4	0	Stable	2.22000	11.4574	28.296
Lithium-6	0	Stable	2.70130	13.7612	31.994
Lithium-7	+1/+1 PEP gap	Stable	2.70130	11.7953	39.245
Lithium-8	+2/+1 PEP gap	839.40 ms	2.98165	13.8793	33.243
Beryllium-8	0	8.19×10^{-17} s	2.98165	13.8793	45.765
Lithium-9	+3/+1 PEP gap	178.3 ms	2.98288	12.3525	34.124
Beryllium-9	+1	Stable	2.90211	11.3760	58.164
Beryllium-10	+2	1.39×106 y	2.99710	11.2770	49.940
Boron-10	0	Stable	3.05071	11.8931	58.800
Boron-11	+1	Stable	3.23034	12.8364	76.205
Boron-12	+2	20.20 ms	3.26036	12.0978	63.240
Carbon-12	0	Stable	2.90211	8.5320	92.162

The lesson to take away here is that a lower MBS volume per proton ratio is representative of the "push for decay," yet structural integrity is another component. These factors are sometimes at odds with one another, resulting in very different half-lives depending on the situation, as described in Table 2.2.

Moving from lithium-9 to beryllium-9 or from boron-12 to carbon-12 provides a better nucleus volume per proton ratio, whereas from beryllium-10 (Fig. 2.8) to boron-10 (Fig. 2.9) it moves in the opposite direction. The negative overcharge of beryllium-10 in relation to the normal 1 to 2 ratio in the nucleus wins over, but only just—beryllium-10 has a half-life of 1.39×10^6 years. The gain in binding energy when arriving at boron-10 might also help with the decay.

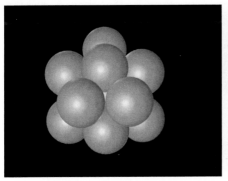

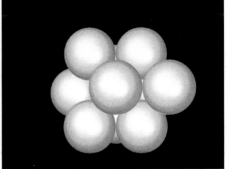

Figure 2.8 Beryllium-10.   **Figure 2.9** Boron-10.

We also recognize the big increase in binding energy with carbon; the icosahedron structure is very dense, as discussed earlier. After β– decay the nucleus changes shape to achieve a better suited structure. We have the same components, but they are arranged in a different pattern, forming a different structure.

2.4 A FIRST LOOK AT THE POSITIONS OF THE INNER ELECTRONS

Inner electrons start with the nucleus of hydrogen-2, also named a deuteron (Fig. 2.10, *left*). This is an important basic building block of the nucleus. We have already seen the positioning of the protons up to carbon-12. Now we will position the inner electrons to create the deuterons in the nucleus up to carbon.

The protons are still marked gray. The electrons are of a light-yellow color and show up in the space between the protons akin to the form of a torus. The electron position in helium-3 is a research topic: Does it stay with the single deuteron or does it move into the center (depicted as such in Fig. 2.10)? Helium-4 is constructed as two deuterons in a tetrahedral configuration.

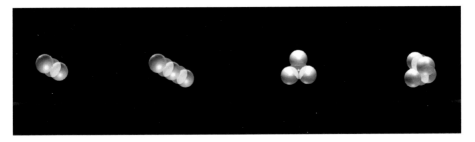

Figure 2.10 Inner electron positions of hydrogen-2, hydrogen-3, helium-3, and helium-4.

Figure 2.11 Proton/neutron positions of hydrogen-2, hydrogen-3, helium-3, and helium-4 in the Standard Model.

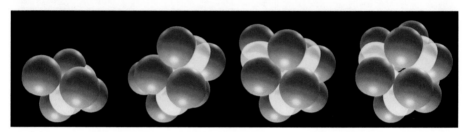

Figure 2.12 Inner electron positions of lithium-7, beryllium-9, boron-10, and carbon-12.

We can compare this setup with the Standard Model. The protons are colored gray, the "neutrons" are marked yellow (Fig. 2.11).

At least in this early stage it is clearly visible how the inner electron takes the role of the electron that is hidden within the "neutron." We continue up to carbon with the setup of the inner electrons (Fig. 2.12). However, the Standard Model cannot provide a built-in attractive force starting from helium-3. This is why the need for the "strong force" arises in it.

Lithium-7 has three deuterons and an extra PEP, beryllium-9 has four deuterons and an extra PEP, boron-10 has five deuterons, and carbon-12 has six deuterons. The carbon icosahedron structure is hollow on the inside, but still the positioning of the inner electrons provides a very stable, dense packing.

The positioning of the inner electrons becomes somewhat arbitrary with heavier nuclei, but we must consider that the electrons also repel each other, so they would spread out as evenly as possible. This limits the number of options.

2.5 METASTABLE ISOTOPES AND ELEMENTS

We can clearly see from the scheme developed here, that a small nucleus tends to perform β– decay when at least two additional PEPs are available in it. This however does not mean the nuclear reaction that follows is instantaneous.

Each nucleus is different, in the sense that an equilibrium is found in the positioning of protons and inner electrons. One nucleus may be able to hold two PEPs completely stable due to the availability of a so-called "PEP gap." Other nuclei may experience almost immediate reactions. Metastable isotopes and elements sit somewhere in between. A good example is the carbon-14 isotope, which has a half-life of

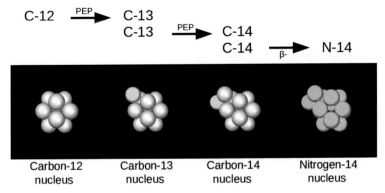

Figure 2.13 Nucleus buildup from carbon to nitrogen.

about 5,730 years and decays to nitrogen-14 (Fig. 2.13). This isotope forms the basis of the radiocarbon-dating method.

In theory there can be multiple locations where an extra PEP would be able to attach to a nucleus. However, many of these theoretical places are not viable in the sense that they would violate the densest packing principle for that particular nucleus.

2.6 DISTORTING THE PERFECT SHAPE

Nitrogen has an icosahedron-like shape with an additional deuteron attached to its side. However, there is more to this. Let us compare carbon-14 (Fig. 2.14) and nitrogen-14 (Fig. 2.15).

One can note that a slight gap between the protons has opened in the center of nitrogen-14 (marked by the light-yellow ring). The additional deuteron on the side (created by β− decay) is pulling the dense icosahedron structure apart, distorting it.

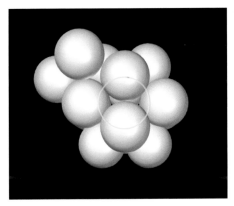

Figure 2.14 Carbon-14 with yellow ring showing no gap.

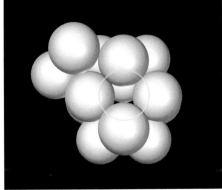

Figure 2.15 Nitrogen-14 with yellow ring showing the gap.

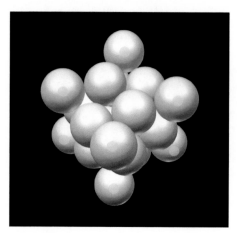

Figure 2.16 Carbon-20, that is carbon-12 with 8 additional PEPs in a cubic setup.

Nitrogen is the first element that has this distinct distorted icosahedron shape. Since β– decay converts the two available PEPs into a deuteron, we therefore move from carbon to the next element: nitrogen. For the deuteron to be created, the icosahedron structure must be distorted. This is a trade-off between the optimal structure of carbon and the creation of the next element through a new deuteron. *Only in this way is growth beyond carbon possible.*

Can we add more than just two PEPs to carbon before β– decay sets in? In theory, we can load carbon-12 with at least eight PEPs to arrive at carbon-20 (Fig. 2.16). These PEPs would be distributed in something of a cubical shape over the structure of the icosahedron.

Interestingly, considering the known isotopes of carbon, carbon-20 is the last isotope observed with PEPs firmly attached. Heavier isotopes are at least partially theoretically derived or show so-called "halo-neutrons." SAM therefore might be showing a strong correlation between the maximum number of isotopes and the maximum number of PEPs we can add to an element. We will look into this topic in Section 4.1. For now, we move from nitrogen to oxygen (Fig. 2.17).

In Figure 2.17 (*far right*) oxygen-16 is depicted as a carbon with two additional deuterons on one side, opening up at the top. The gap in the basic icosahedron structure is still present. This is the trade-off for allowing structures to be created resembling tetrahedrons.

N-14 $\xrightarrow{\text{PEP}}$ N-15

N-15 $\xrightarrow{\text{PEP}}$ N-16

N-16 $\longrightarrow$ C-12

N-16 $\xrightarrow[\beta-]{\beta-\ \alpha}$ O-16

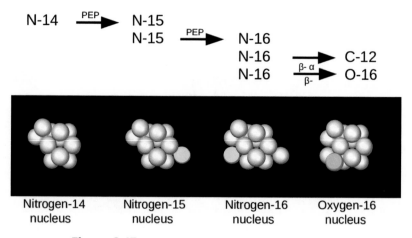

| Nitrogen-14 nucleus | Nitrogen-15 nucleus | Nitrogen-16 nucleus | Oxygen-16 nucleus |

Figure 2.17 Nucleus buildup from nitrogen to oxygen.

2.6.1 A little bit of SAM history

When in the early phase of this research the connection between shape and the first elements was made, a problem emerged. As we now know, the icosahedron represents carbon. The icosahedron is the perfect spherical dense packing for 12 spheres—many researchers give this structure some credence. However, when we use spheres in creating larger and larger structures in a geometric fashion, the problem that arises is that we cannot really build beyond the icosahedron (carbon). There is no larger geometric structure based on platonic solids than the icosahedron and any addition to the structure leads automatically to limitations. The tetrahedron densest-packing principle (any addition must complete a tetrahedron) breaks the icosahedron structure when we keep adding, as small gaps appear between the spheres that do not have enough room for sphere placement, nor are they touching.

The alpha particle model (see Section 7.1) for example creates geometric structures exclusively with the tetrahedron (a neon-20 would consist of 5 tetrahedrons or α particles or helium-4 nuclei). However, we see a problem with larger structures because the geometry again becomes too large to keep the tetrahedron pattern intact.

Many models in nuclear physics deal with the spatial component in their own way. Some try to use geometric shapes within geometric shapes (shell models) but do not show strong correlations with the known properties and often too many arbitrary choices have to be made. Others avoid the problem of geometric structures by offering a lattice model in a layered system. However, this leads in turn to arbitrary choices and too many options again.

The following is a personal account by Edo Kaal, remembering the early days of developing the model:

When using the neodymium di-polar magnets to see how the shapes would act, I learned some noteworthy lessons. First, the magnets—when properly handled—can be made to precisely mimic the structure of the smaller nuclei. Helium-4 is almost impossible to make with the magnets, but otherwise they all work.

Then the issue of elements larger than carbon was contemplated. The icosahedron as shown in Figure 2.18 (top right) was a beautiful representation of the densest packing scheme. However, no matter what I tried, the larger structures always broke or distorted the underlying carbon icosahedron structure. For months, I tried to find another angle of solving this issue and finally gave up and concluded, that I was at an impasse with the model. After a few weeks, I picked the magnets up again and tried to go back to the roots of science: observation!

I recognized that larger structures cannot be made with platonic solids!

However, who said this has to be the case? After I decided to accept distortion as necessary, I found the key for the larger structures. The moment we add two magnets (representing a deuteron for the next element step) next to each other on a carbon base built from magnets we see that the carbon base is distorted by a gap that opens between two normally connecting magnets. This does not happen when two extra magnets are added separately at different positions, as occurs with single PEPs.

Figure 2.18 Structures of neon-20, carbon-12, lithium-7, and helium-4 created using magnets.

What the magnets taught me was that I was ignoring the forces at play. We cannot make a change to a system without keeping this in mind. We cannot act as if we add a proton–electron pair to a nucleus and nothing happens.

The magnets showed me that nitrogen has a gap in the carbon base (Fig. 2.19). The gap exists to accommodate the extra protons in the nucleus to make three perfect tetrahedrons which we now know as a capping deuteron. The creation of that deuteron on top of the icosahedron opens the carbon and primes the places for the other three capping

Figure 2.19 Nitrogen-14 created from magnets with visible gap.

deuterons to take their place. This finding happened due to 'playing with magnets.' The magnets became in a way a great teaching tool for me.

This finding showed me furthermore that we need to deal with the geometry in a different manner, namely, to integrate additional icosahedrons side by side— creating the backbone—as proposed in the model. Nature solved the problem of growth this way, and that way is reflected in the elements in the periodic table.

In conclusion, the magnets taught me that the broken geometry and the error that occurs is not an error at all. It became clear that the gap was essential, crucial for understanding larger elements. This is now a part of SAM."

2.7 BALANCING THE NUCLEUS

To reach oxygen-17 from oxygen-16 a PEP is added to the side (*right*) that is not covered with the V-shaped structure (Fig. 2.20). A second PEP goes to the other (*left*) side of the nucleus, filling the V-shaped structure, and we arrive at oxygen-18 (Fig. 2.20, *center*). The pattern resulting from adding additional PEPs is continuing. Therefore, we would expect oxygen-18 to be unstable as it already carries two additional PEPs compared with oxygen-16. However, it is stable. Oxygen-19 is the first unstable isotope of oxygen.

Is there a structure in oxygen ready to receive a PEP, similar to the case of lithium-6? Indeed, it is the V-shape created by the two deuterons, depicted in green for better visibility (Fig. 2.21)—a place that would readily accept another PEP.

When we look at oxygen-18 in Figure 2.22 we can still see the V-shaped structure, but it is now filled with a PEP. This step did not adversely affect the stability of the nucleus. There are similarities visible with lithium-6 (Fig. 2.23). We can add another PEP to the right side of the nucleus of oxygen-18, to reach oxygen-19 (Fig. 2.20). This is very similar to the step from lithium-6 to lithium-9. The gap allowed for one more PEP to be added, before the nucleus reacts with β– decay, and we end up with fluorine-19 (Fig. 2.20, *right*).

O-16 $\xrightarrow{\text{PEP}}$ O-17

O-17 $\xrightarrow{\text{PEP}}$ O-18

O-18 $\xrightarrow{\text{PEP}}$ O-19

O-19 $\xrightarrow{\text{β-}}$ F-19

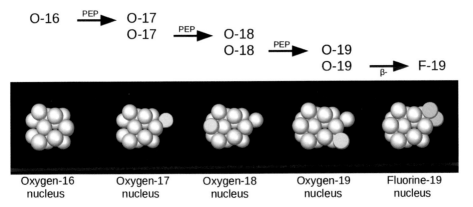

| Oxygen-16 nucleus | Oxygen-17 nucleus | Oxygen-18 nucleus | Oxygen-19 nucleus | Fluorine-19 nucleus |

Figure 2.20 Nucleus buildup from oxygen to fluorine.

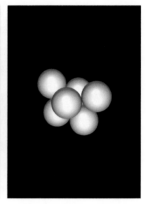

Figure 2.21 V-shaped structure of oxygen-16.

Figure 2.22 V-shaped structure and PEP of oxygen-18.

Figure 2.23 Lithium-6.

We made it a structural rule in SAM that the 5th proton on the V-shaped structure (Fig. 2.21)—making it a V-shaped structure with an additional PEP (Fig. 2.22)—is required before the 3rd deuteron building block can be placed on the opposing side of the nucleus.

Is there a stable fluorine-18 without the additional PEP to the left? Apparently not, since fluorine-19 has an isotopic abundance of 100%, hence there is only this one isotope. This could be driven by the fact that the deuterons on one side of the icosahedron structure show this so-called "PEP gap," a phenomenon that we know from observation, reflecting the tendency of a nucleus to absorb a PEP.

Fluorine-18—if artificially created—is unstable. It decays with a half-life of 110 minutes to oxygen-18 by $\beta+$ decay or electron capture.

2.8 STRUCTURALLY REQUIRED EXTRA PEPS

With fluorine-19 we have the first case of a structurally required extra PEP. With lithium-6 the extra PEP (to reach lithium-7) was not required, just preferred, but here with fluorine-19 in 100% abundance it is required to maintain structural integrity. Arguably the first structurally required PEP manifests at beryllium-9. However, beryllium-9 is unique and not a repeating case, as is the case with the configuration of fluorine-19. Adding one further PEP to fluorine-19 derives first fluorine-20 and then neon-20 after a $\beta-$ decay, a noble gas. The rearrangement moves a proton from the PEP gap on the left side over to the right side as part of the newly created deuteron (Fig. 2.24).

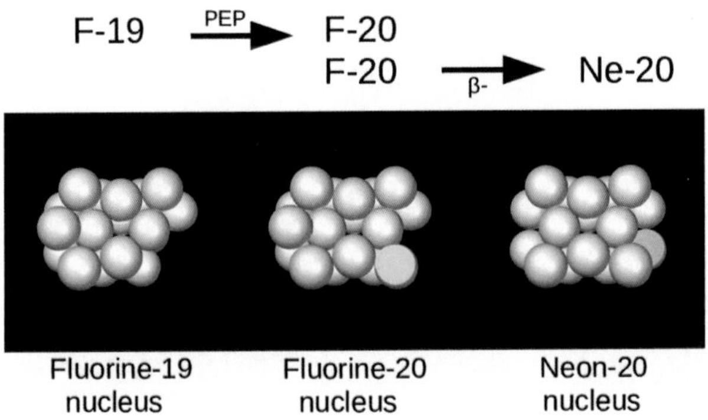

F-19 $\xrightarrow{\text{PEP}}$ F-20

F-20 $\xrightarrow{\beta-}$ Ne-20

| Fluorine-19 | Fluorine-20 | Neon-20 |
| nucleus | nucleus | nucleus |

Figure 2.24 Nucleus buildup from fluorine to neon.

It is at this step when a balanced state with deuterons is reached and that occupying the PEP gap is no longer structurally required.

2.9 SOME THOUGHTS ON NOBILITY

The growth pattern from carbon to neon is now clear—effectively adding four deuterons to the icosahedron structure (i.e., two on each side). However, the process is not as simple as adding deuterons or PEPs in the right place since multiple steps are required to balance the nucleus and then rearrange it to reach the four-deuteron-topped icosahedron stage. Neon-20 has four complete deuterons attached to the icosahedron structure.

In terms of geometry, a tetrahedron has 4 triangular faces and an icosahedron has 20 triangular faces.

Helium-4, which is composed of two deuterons in a tetrahedron shape, has two inner and two outer electrons. The icosahedron (the carbon-like structure) has 12 protons, six inner electrons, and six outer electrons. Neon-20 also has a basic icosahedron structure with 20 faces and four additional deuterons, and therefore supports 10 inner and 10 outer electrons. In a way we can say that all faces of the basic structure are covered, either by inner or outer electrons: four faces in the case of helium-4 and twenty faces for neon-20. Is this the reason some elements are "noble" (i.e., more or less chemically inert)? We will have to see if this pattern holds for other cases. What we can conclude here is that the nobility or inert property of certain elements is an effect of all the electrons at play—inner as well as the outer. This also raises the question whether chemistry is really only about outer electrons, or whether there is more to this? We should recall that one of the goals of SAM is to bring nuclear physics and chemistry back together. We will consider this topic further in Sections 2.15, 4.5, 5.4, and 6.1.

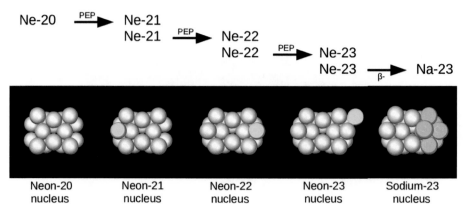

Ne-20 $\xrightarrow{\text{PEP}}$ Ne-21
Ne-21 $\xrightarrow{\text{PEP}}$ Ne-22
Ne-22 $\xrightarrow{\text{PEP}}$ Ne-23
Ne-23 $\xrightarrow{\beta-}$ Na-23

| Neon-20 nucleus | Neon-21 nucleus | Neon-22 nucleus | Neon-23 nucleus | Sodium-23 nucleus |

Figure 2.25 Nucleus buildup from neon to sodium.

2.10 GROWTH BEYOND THE NOBLE STATE

How do we get growth from a noble state onward? First, there is no reason to assume that a noble element cannot capture additional PEPs, especially since this would not change the setup that makes them noble.

We notice there are two possible spots on the nucleus that can and will work as PEP gaps. Arbitrarily the left PEP gap is filled first (neon-21), then the new one on the right. Neon-22 is still stable, making up 9.25% of available neon.

Neon-23 is no longer stable since with β– decay it rearranges to sodium-23 (Fig. 2.25). On the left side sodium-23 looks like lithium, depicted in red in Figure 2.26. The lithium-like structure is attached to the inner, carbon-like structure (icosahedron), while the left side is still covered by an additional PEP in the V-shaped structure created from two deuterons to balance against the other side. The number of protons involved is 12 (icosahedron) + 6 (lithium-like) + 4 (V-shaped structure) + 1 (PEP) = 23.

The isotope sodium-23 makes up 100% of available sodium with sodium-22 and sodium-24 only found in trace amounts—because they are so unstable. The PEP on the V-shaped structure seems to be again structurally required—to balance the lithium-like structure on the other side. Sodium-22 is missing this PEP and decays by β+ decay or electron capture to neon-22 after 2.6018 years.

When moving from sodium to magnesium-24, the new element has two

Figure 2.26 Lithium-like structure (*red*) as part of sodium.

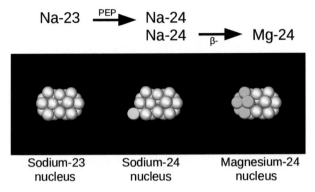

Na-23 $\xrightarrow{\text{PEP}}$ Na-24
Na-24 $\xrightarrow[\beta-]{}$ Mg-24

| Sodium-23 | Sodium-24 | Magnesium-24 |
| nucleus | nucleus | nucleus |

Figure 2.27 Nucleus buildup from sodium to magnesium.

lithium-like structures, one on each side (Fig. 2.27). Growth on the V-shaped structure with an additional PEP is preferred to growth on the lithium-like structure. It might also be related to the ongoing balancing act needed to build up the nucleus.

2.11 MOVING FROM NUCLEUS TO NUCLEUS

To complete the second row in the periodic table of elements we move on up to argon.

Magnesium is stable up to magnesium-26. With magnesium-27 another β– decay sets in, followed by a rearrangement to aluminum-27 (Fig. 2.28, *right*). The V-shaped structure with a filled PEP gap is back on the left side—last seen with sodium-23, the right side showing a structure like boron with 10 protons.

With silicon-28, we see the V-shaped structure with an additional PEP on one side and on the other we now have a structure like carbon-12 (sharing a proton with the inner carbon-like structure—an icosahedron) (Fig. 2.29).

When moving from silicon-28 to phosphorus the left side of phosphorus-31 has a beryllium-like structure; on the right is the recognizable carbon-12-like structure (Fig. 2.30). One question arises at this point: Why is there no combination of a lithium-like structure and a carbon-like structure? Why do we go directly to the beryllium-like structure? We will discuss this in Section 4.4.

With sulfur-32 the V-shaped structure with an additional PEP on the left side is back—to the right we have another carbon-like structure with another V-shaped structure (Fig. 2.31) on top of it. Phosphorus-32 is created here with PEP capture from phosphorus-31. Phosphorus-32 has important uses in medicine, biochemistry, and molecular biology. It only exists naturally on Earth in very small amounts and its short half-life means useful quantities have to be produced synthetically. Phosphorus-32 can be generated synthetically by irradiation of sulfur-32 with so-called moderately fast-moving PEPs.

Chlorine-35 (Fig. 2.32) is very similar to sulfur, but it carries an additional PEP on the added carbon and one added deuteron. The right side looks like fluorine-19, which belongs to the same group.

Mg-24 $\xrightarrow{\text{PEP}}$ Mg-25
         Mg-25 $\xrightarrow{\text{PEP}}$ Mg-26
                  Mg-26 $\xrightarrow{\text{PEP}}$ Mg-27
                           Mg-27 $\xrightarrow{\beta-}$ Al-27

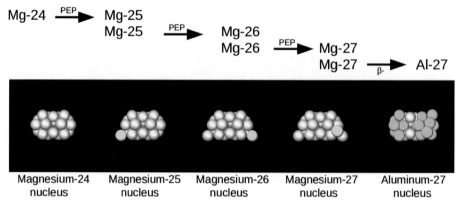

| Magnesium-24 | Magnesium-25 | Magnesium-26 | Magnesium-27 | Aluminum-27 |
| nucleus | nucleus | nucleus | nucleus | nucleus |

Figure 2.28 Nucleus buildup from magnesium to aluminum.

Al-27 $\xrightarrow{\text{PEP}}$ Al-28
         Al-28 $\xrightarrow{\beta-}$ Si-28

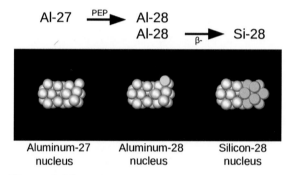

| Aluminum-27 | Aluminum-28 | Silicon-28 |
| nucleus | nucleus | nucleus |

Figure 2.29 Nucleus buildup from aluminum to silicon.

Si-28 $\xrightarrow{\text{PEP}}$ Si-29
         Si-29 $\xrightarrow{\text{PEP}}$ Si-30
                  Si-30 $\xrightarrow{\text{PEP}}$ Si-31
                           Si-31 $\xrightarrow{\beta-}$ P-31

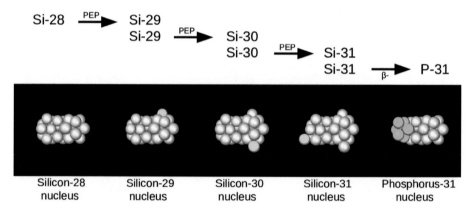

| Silicon-28 | Silicon-29 | Silicon-30 | Silicon-31 | Phosphorus-31 |
| nucleus | nucleus | nucleus | nucleus | nucleus |

Figure 2.30 Nucleus buildup from silicon to phosphorus.

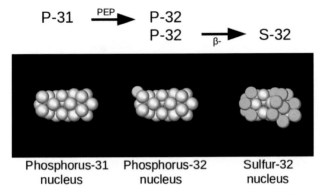

Figure 2.31 Nucleus buildup from phosphorus to sulfur.

The growth process is starting to get complex, with three growth points simultaneously. On our way to argon we would once again expect a simple step from chlorine-35 to chlorine-36 to argon-36, but the available abundance data clearly indicates that argon-40 is the dominant isotope—with an abundance on Earth of 99.6035%. Argon-36 is clearly not the isotope we need to reach, it is argon-40. The new third growth point on the nucleus now comes into play. The path to argon-36 and onward to argon-38 involves a two-PEP step in one go to avoid unstable argon-37 with its half-life of 35 days. Chlorine-36 is metastable with a decay time of 3.01×10^5 years. It starts to build a deuteron for the well-known V-shaped structure and then decays to argon-36 by β– decay. Argon-36 has two V-shaped structures that are ready to receive additional PEPs, which are readily filled to reach argon-38 (Fig. 2.33).

However, chlorine-37 must also be considered as an alternative since it makes up 24.24% of the available chlorine on Earth. It can be created from sulfur-36, which is also stable, but only exists in trace amounts (Fig. 2.34). Sulfur-36 is created from chlorine-36 by β+ decay (Fig. 2.33).

We still have not reached argon-40. Argon-39 is in our way, it is unstable with a half-life of 269 years and it exists only in trace amounts. Is there another path? Let's follow where argon-39 takes us (Fig. 2.35).

Potassium-39 sees the return of a lithium-like structure, as happened with sodium before. "Coincidentally" they belong to the same group.

Nucleus development gets even more interesting as we progress to potassium-40. Here, in addition to the normal path moving on to calcium-40, in at least 10% of decay cases we move downward to argon-40 by β+ decay, which is the most abundant form of argon on Earth, at 99.604% (Fig. 2.36).

Calcium-40 presents us with two lithium-like structures on top of the second carbon-like structure. Magnesium shows the same configuration on the initial carbon structure—and again they belong to the same group.

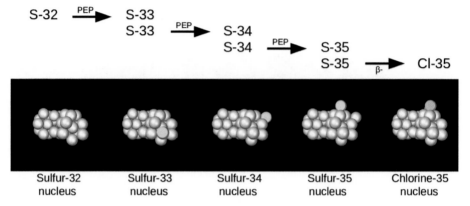

Figure 2.32 Nucleus buildup from sulfur to chlorine.

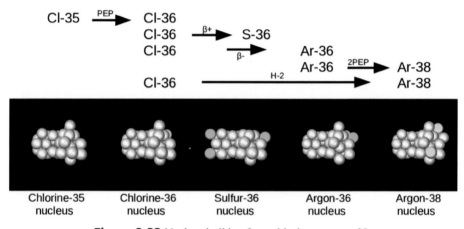

Figure 2.33 Nucleus buildup from chlorine to argon-38.

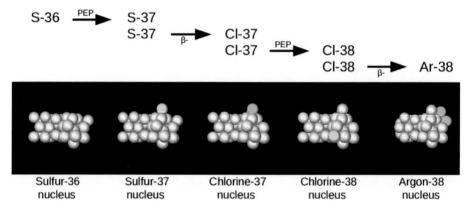

Figure 2.34 Nucleus buildup from sulfur-36 to argon-38.

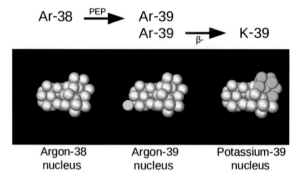

Figure 2.35 Nucleus buildup from argon-38 to potassium-39.

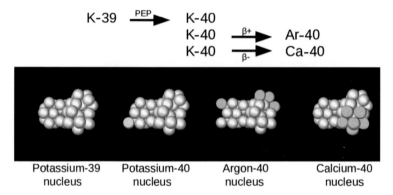

Figure 2.36 Nucleus buildup from potassium-39 to argon-40 and calcium-40.

2.12 GROWTH PATTERNS OF THE OVERALL STRUCTURE

2.12.1 Building blocks

We have seen how structure governs and directs growth through the rules of spherical dense packing—moving from element to element and moving into variations of elements (isotopes). We did this, for the most part, by adding PEPs to a nucleus and correcting the number of electrons in the nucleus by β– decay. We have created nuclei up to calcium in our model and already noticed that structural patterns are emerging—recurring sub-structures on top of carbon-like structures. Let's call them *building blocks* for now.

Nucleus of deuterium: Two protons (one deuteron)
   V-shaped structure: Two deuterons in a V-shaped structure; comparable with a squashed tetrahedron

Nucleus of lithium: A ring of five protons with a cap proton on one side of the ring packed into three deuterons; the upper part of the pentagonal bi-pyramid

Nucleus of beryllium: Four deuterons packed together

Nucleus of boron: Five deuterons packed together

Nucleus of carbon: Two rings of five protons with a cap proton on top of the upper one and on the bottom of the lower one (icosahedron); or six deuterons packed together in two rings with caps.

Figure 2.37 shows all the possible building blocks. They are listed below in the order in which they grow. Also, we present a color scheme that highlights the various building blocks. This color scheme will be adopted throughout the book from now on but only signifies structure, not material difference. All are just protons, maybe sometimes paired with an electron. Additionally, a single PEP will be marked yellow and a single proton will be marked light brown:

1   *Two capping*—when the icosahedron is growing and just one deuteron is attached to the carbon-like structure. We call this a *two-ending* (*green*).
2   *Neutral capping*—when a branch has grown to a neutral point and is no longer chemically reactive. Four additional protons and two inner electrons have been added to one side of a carbon-like structure in a V shape. We call this a *four-ending* (*green*).
3   *Neutral capping/four-ending with an additional PEP*—an additional PEP is added to balance the structure or create an isotope. We call this a *five-ending* (*green* with a *yellow* PEP).
4   *Lithium nuclet* (*six protons*)—the next step in growth following neutral capping and the addition of a PEP is to add another proton and create a so-called *lithium nuclet* (*red*).
5   *Beryllium ending* (*eight protons*)—this is an intermediate state between lithium and carbon nuclets (*orange*).
6   *Boron ending* (*ten protons*)—this is also an intermediate state (*purple*).
7   *Carbon nuclet* (*eleven protons*)—this is a completed icosahedron (*blue*).
8   *Initial ending*—both sides of the carbon are capped making this branch completely chemically inert. When all branches are capped the element is noble (*light-blue* or *white*).
9   *Final ending*—the final shape of the carbon nuclet representing the so-called "backbone" of the nucleus. We call this a *backbone nuclet* (*white*).

We coined the term "nuclet" ourselves. What we mean by it will become clear in Section 2.12.3, as well as the meaning of "ending" and "capping."

There is also a difference between the initial carbon and the following carbons that connect to it. The connection happens through a shared proton, which means that subsequent carbons in the backbone provide only 11 protons to the structure—not 12. As a consequence a new secondary carbon in the backbone contains only 5 deuterons and a single proton—not 6 deuterons.

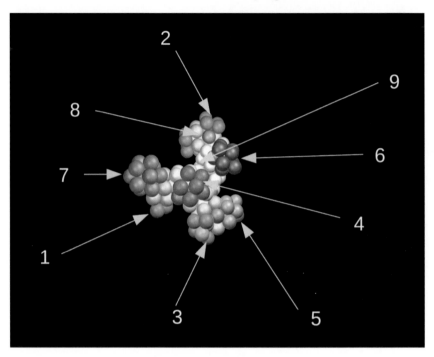

Figure 2.37 Fictional nucleus showing all endings, cappings, and nuclets (building blocks).

2.12.2 Phases

For the first three rows of the PTE, growing phases of endings are clearly visible. Starting from the icosahedron (carbon-12) we first see the capping phase. In it we see growth from the two-ending over the four-ending to the five-ending. The completion of the capping phase gives us a noble gas (neon). Then we switch to the building phase. In this phase we start with the lithium ending, followed by the beryllium ending, the boron ending and finally end with the full icosahedron of the carbon ending. An ending grows as a general rule with steps of one deuteron (d)—sometimes only a single proton (single p)—switching between *building and capping* phases:

	1 d + single p	up to	2 d + single p	exceptions (capping phase)
>	2 d + single p	up to	6 d + single p	exceptions (building phase)
>	6 d + single p	up to	10 d + single p	capping phase
>	10 d + single p	up to	14 d + single p	building phase
>	14 d + single p	up to	18 d + single p	capping phase

The core carbon structure of neon-20—the result of the first real capping phase—is the first piece of the nucleus "backbone." It is the primary building block of larger nuclei. These are combined in a tree-like fashion with this structure, creating the backbone of the nucleus. It is the endings of the branches of this structure that are different and active.

The eight elements in the 2nd row of the periodic table of elements (i.e., lithium, beryllium, boron, carbon, nitrogen, oxygen, fluorine, and neon) are each composed of a single base ending in various stages of growth. After carbon is reached two new growth points emerge which are being capped. Neon, the last element in the row, is inert because its carbon core is completely capped.

The eight elements of the 3rd row of the periodic table of elements (i.e., sodium through argon) start and complete the growth of the new carbon nuclet to a completely capped state. The last element in the row, argon, is inert because it consists of two completed (completely capped) backbone nuclets.

After this point we see parallel growth on several more growth points, which make the phases less visible, but they still hold up for each ending. However, some steps in the building and capping phase become less common than others.

2.12.3 Definitions

Above we used three terms in the description of Figure 2.37 and in Section 2.12.2 that we have not yet defined properly.

Definition

An *ending* is a stable, geometrically arranged cluster of protons and inner electrons as defined in Section 2.12.1 as building block.

We discussed the phases of the growth in Section 2.12.2. We can safely talk about endings starting from nitrogen upward. The two-ending, four-ending, and five-ending belong to the capping phase. The other endings belong to the building phase.

Definition

A *nuclet* is a densely packed, geometrically arranged cluster of protons and inner electrons that belong to the building phase. This is an invented word. Do not try to look it up since you will not find it.

Therefore, we can now talk about the lithium, beryllium, boron, and carbon nuclet. We can still call them "endings" too.

Definition

A *capping* is a geometrically arranged cluster of protons and inner electrons that belong to the capping phase.

Therefore, we can now talk about the two-, four-, and five-capping. We can still call them "endings" too.

2.12.4 Conventions

Looking at the description of Figure 2.37 we notice that we only use the nuclet name for carbon and lithium. Why not use it for the beryllium ending and boron ending since the definition would allow it? The carbon structure is special anyway, and so is lithium. Why lithium? Because beryllium-9 can be viewed as two lithium-7s intersecting and boron-11 can be viewed as three lithium-7s intersecting (Fig. 2.38). Carbon-12, by the way, can be viewed as four lithium-7s intersecting. From this vantage point it makes sense that magnesium is in the same group as beryllium, but it does not carry a beryllium ending. Instead, it carries two lithium nuclets. That is the same thing.

2.12.5 The big picture

Considering the big picture, it appears that after carbon, the nucleus grows by adding predefined sub-clusters to its own structure. The building of the nucleus apparently follows a blueprint. However, at this stage we can already see the beginnings of a tree-like structure emerging as growth points in the structure develop into further carbon nuclets with active endings.

2.13 RETHINKING ELEMENT CREATION

With some early exceptions we used the scheme of adding PEPs to the nucleus until it becomes unstable before applying β– decay to change it to the next element as a means of building the elements. However, as already detailed, sulfur, chlorine, and argon form pivotal points in the scheme. Before this point there was a continuous block of stable isotopes per element that allowed for easy steps—adding two or three PEPs— resulting finally in a β– decay and changing the element number upward by one. However, from this point on we can see that this is no longer enough, as there are un-stable isotopes interspersed between stable ones. We need to add more PEPs in a more complicated scheme before the β– decay can happen. With chlorine this became

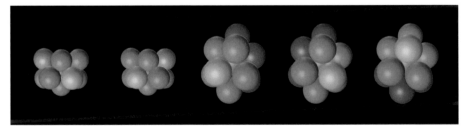

Figure 2.38 Beryllium viewed as two intertwined lithium nuclets (*left*); boron viewed as three intertwined lithium nuclets (*right*).

more difficult and with argon we had to use β+ decay to reach argon-40—the most predominant isotope of argon on Earth. Additionally, when considering element abundance in the Sun, we note that argon-36 can be identified in its spectrum but argon-40 cannot. Where does this difference come from? How does nature actually do element creation?

> The Standard Model currently states that elements are created *inside stars only.* After reading this book, the reader might be willing to accept that there is a much more detailed and complex means by which elements are created—not only in stars.

Can fusion happen in bigger chunks, as with helium or oxygen for example? Is sulfur-32 possibly created through an oxygen–oxygen fusion? Does fusion really stop at iron to be energetically preferred? We will try to answer these questions in Chapter 12.

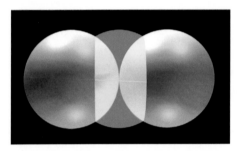

Figure 2.39 Hydrogen-2 nucleus with a "torus-shaped" inner electron of proton size.

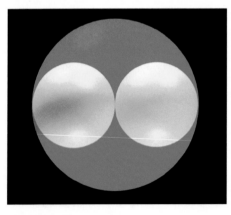

Figure 2.40 Hydrogen-2 with inner electron influence twice the size of a proton.

2.14 INNER ELECTRONS

It is now time to look at electrons in more detail. Let us start with the inner electrons. We already looked at their positions for the first elements in Section 2.4. The inner electrons are considered the glue keeping the protons together in the nucleus. The typical minimal configuration is the deuteron, with two protons and one electron. The natural position of the electron would be between the two protons—holding them together with electrostatic force. An inner electron needs at least two anchor points to be stable, in this case the two protons. Assuming the protons would touch, the electron between them should have the form of a torus (Fig. 2.39). However, the size of the inner electrons appears to be bigger than the proton size. One inner electron can cover the range of several protons—at least if we consider it to be a sphere of influence and not necessarily a physical particle presence (Fig. 2.40).

The inner electrons are in fixed positions per element and there is no reason they would shift or relocate. If they do, that would be a nuclear reaction or the initiation of a nuclear reaction. Collectively, in the carbon nuclet they will therefore be positioned in an octahedron shape which is again a perfect platonic solid. The size of the electrons as well as their position creates a "sphere" that glues the protons in the icosahedron together while creating spots of more or less overall positive charge distributed over the surface of the icosahedron (Fig. 2.41).

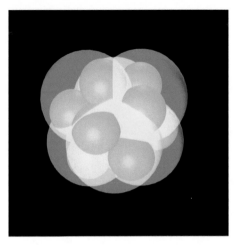

Figure 2.41 Carbon-12 with inner electron influence twice the size of a proton.

To get a better grasp of the structure the protons are displayed in Figure 2.42 as gray and small, while the electrons appear larger and yellow. We see the deuteron configuration of hydrogen-2 and helium-4 (which is represented by two deuterons perpendicular to each other, creating the tetrahedron structure—green wire frame—of the protons). With carbon, we see the icosahedron structure—the blue wire frame—of the protons and the inner octahedron—yellow wire frame—of the six electrons (Fig. 2.42).

We already stated (Section 2.12.1) that the core of the nucleus is made up of deuterons and possibly a few single protons, but there are also additional PEPs which can connect through the electron to the structure if, as a consequence of the connection, a tetrahedron structure is completed. The next connecting PEP might be able to create another deuteron. This would mean one too many electrons in the area, with an electron being emitted from the nucleus via β– decay.

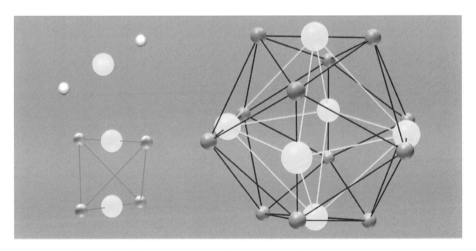

Figure 2.42 Inner structure of hydrogen-2, helium-4, and carbon-12.

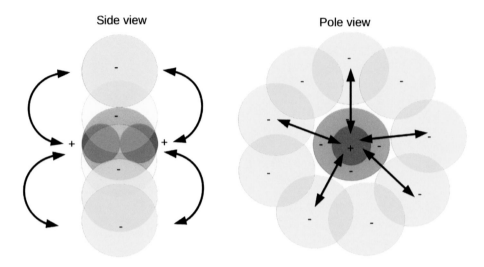

Side view   Pole view

Figure 2.43 The position of the hydrogen-2 outer electron (*light yellow*). The protons are displayed in *blue*, the inner electron in *orange*.

2.15 OUTER ELECTRONS

We again start with the deuteron nucleus when looking at outer electrons. This nucleus has two positive spots at the outer ends of the protons on both sides—the poles—while the inner electron sits in the middle. With deuterium (hydrogen-2) there is one outer electron. Outer electrons, even at a distance, connect to where the positive points are on the nucleus. With the deuterium atom, it is obvious where that is, the outer electron will be repelled by the inner electron, not having a fixed position and having to choose between two equal spots on opposite sides. It would take a position as far as possible away from the inner electron, yet as close as possible to the still positive nucleus on both ends. Probably it will settle between the two connection points, rotating around the deuteron nucleus in an imaginary *z*-axis in and out, as displayed in Figure 2.43.

With hydrogen-1 there is no distinctive positive spot or even plane, the outer electron can be anywhere around the proton (Fig. 2.44).

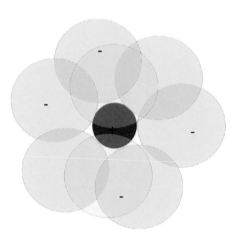

Figure 2.44 Position of the hydrogen-1 outer electron.

There is however an interesting question here: Why does the outer electron (negative charge) not immediately join with the proton (positive charge) to form a PEP? What prevents this from happening? A first clue can be found in Section 7.1 in Table 7.6.

In other, heavier and more complex atoms the outer electrons can, and most likely will, shift positions only a little. Overall, the positions the outer electrons can connect to, and therefore their positions in relation to the nucleus, are fixed. With carbon, for example, electrons will arrange themselves around its icosahedron structure, showing equal spacing. Neon will have a slightly different structure than carbon because of the distortion of the icosahedron and the additional "squashed" tetrahedrons on each side (Fig. 2.45).

In short, the outer electrons arrange themselves in relation to the positive spots on the nucleus (attraction) while distancing themselves from one another due to repulsion. The system balances out the forces. The outer electron structure of carbon and neon would develop a structure based on the dodecahedron because of this balancing act (Fig. 2.46).

Also, outer electrons are more easily movable due to their relatively large distance from the nucleus. As mentioned earlier they will evenly space out between each other and the positive connection points of the nucleus. An outer electron has two anchor points directed toward the nucleus, connecting to two positive spots. An outer electron with more energy (excited state) will be further away from the nucleus, its sphere of

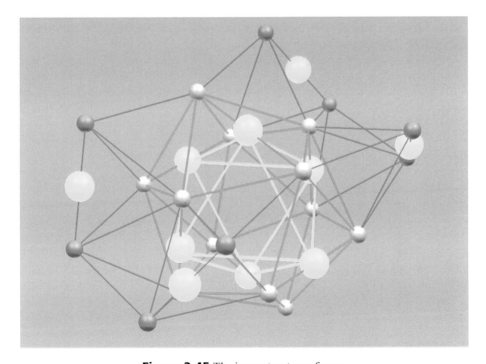

Figure 2.45 The inner structure of neon.

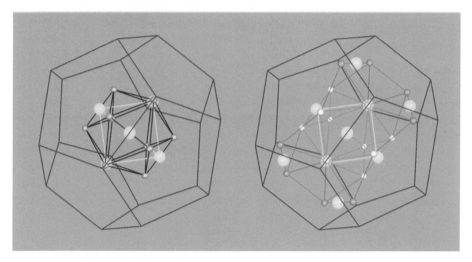

Figure 2.46 Possible carbon and neon outer electron structure.

influence might also be bigger. Since the nucleus will continue to attract the outer electron toward its original position (densest packing again), it will fall back to its ground
state and emit the energy it had absorbed. The reverse of this situation is where extra
energy, perhaps in the form of a photon, is absorbed by an electron, moving it to an
excited state, probably also to a different, more distant outer position.

How exactly the inner structure of the nucleus correlates with the outer electron
structure is still an ongoing research topic for the SAM team. We can describe the
correlation, but the topic is currently just too complex to write a definitive chapter
about.

2.16 THOUGHTS ON THE NUMBER OF INNER ELECTRONS

You would think that SAM equates the number of inner electrons with what is conventionally known as the number of "neutrons." It is also known that conventionally
heavier elements tend to have more "neutrons" (PEPs) than protons. In SAM, we do
the opposite: subsequent carbon nuclets—docking onto the first carbon nuclet forming
the "backbone" of the nucleus—have always 11 extra protons (not 12!) and carry as
a nuclet 5 deuterons and a single proton. Therefore, there are more and more single
protons in the nucleus, as it grows.

Every subsequent final nuclet has this ratio—arriving at iron-54 we find 26 deuterons and 2 single protons, equal to 28 protons and 26 neutrons in the Standard
Model. This is just the other way around, the Standard Model sees 28 neutrons and
26 protons. Is SAM wrong? What needs to happen to "fix" the situation and convert
some protons to PEPs (or kind of a PEP) will be the topic of Section 4.4. There are

huge consequences based on the backbone scheme which will be revealed throughout the book.

2.17 β– AND β+ DECAY STEPS IN GREATER DETAIL

As stated before, we employed β– decay and β+ decay during our buildup of the first elements. As discussed early on in the book (Section 1.5), β decay is the movement of electrons in and out of the nucleus. This now warrants a closer look.

2.17.1 β– decay

First, let us consider all the steps employed so far for a β– decay moving from isotope to isotope.

Isotope	Additional PEPs	Half-life	MBS radius	MBS vol./n#	BE (MeV)
Hydrogen-3	+1	12.32 y	3.00000	37.6991	8.4820
Helium-3	−1	Stable	2.15023	13.8810	7.7180

It is somewhat strange that hydrogen-3 has such a high binding energy, despite not being very densely packed. The much denser packed structure of helium-3 might be the reason the decay happens, despite the decrease in binding energy.

Isotope	Additional PEPs	Half-life	MBS radius	MBS vol./n#	BE (MeV)
Lithium-8	+2/+1 PEP gap	839.40 ms	2.98165	13.8793	33.243
Beryllium-8	0	8.19×10^{-17} s	2.98165	13.8793	45.765

This is not worth too much consideration since beryllium-8 immediately decays into two α particles.

Isotope	Additional PEPs	Half-life	MBS radius	MBS vol./n#	BE (MeV)
Lithium-9	+3/+1 PEP gap	178.3 ms	2.98288	12.3525	34.124
Beryllium-9	+1	Stable	2.90211	11.3760	58.164

This is the first "normal" β– decay—the resulting structure is more densely packed and the binding energy goes up, which means energy is released during the process. This is what we call a normal decay case. In a conflicting case, binding energy goes up and the structure is less densely packed or vice versa.

Isotope	Additional PEPs	Half-life	MBS radius	MBS vol./#p	BE (MeV)
Beryllium-10	+2	1.39×10^6 y	2.99710	11.2770	49.940
Boron-10	0	Stable	3.05071	11.8931	58.800

Here the binding energy goes up, but the resulting structure is less densely packed. This is what we call a conflicting decay case. As a result of the conflict the half-life of the source is (very) high, it resists.

Isotope	Additional PEPs	Half-life	MBS radius	MBS vol./#p	BE (MeV)
Boron-12	+2	20.20 ms	3.26036	12.0978	63.24
Carbon-12	0	Stable	2.90211	8.53199	92.16

This is a normal case—the resulting structure is more densely packed and the binding energy goes up.

Isotope	Additional PEPs	Half-life	MBS radius	MBS vol./#p	BE (MeV)
Carbon-14	+2	5,730 y	3.51367	12.9791	105.28
Nitrogen-14	0	Stable	3.46438	12.4405	104.66

Another conflicting case with a high half-life as expected.

Isotope	Additional PEPs	Half-life	MBS radius	MBS vol./#p	BE (MeV)
Nitrogen-16	+2	7.13 s	4.02432	17.0626	117.98
Oxygen-16	0	Stable	3.55013	11.7139	127.62

This is a normal case.

Isotope	Additional PEPs	Half-life	MBS radius	MBS vol./#p	BE (MeV)
Oxygen-19	+3/+1 PEP gap	26.470 s	4.06499	14.8086	143.76
Fluorine-19	+1/+1 Req.	Stable	4.06499	14.8086	147.8

Close to the normal case with no change in dense packing. The "Req." marking in the additional PEPs column denotes the portion of PEPs that are structurally required PEPs.

Isotope	Additional PEPs	Half-life	MBS radius	MBS vol./#p	BE (MeV)
Fluorine-20	+2/+1 Req.	11.163 s	4.06499	14.0681	154.4
Neon-20	0	Stable	4.02432	13.6501	160.65

This is a normal case.

Isotope	Additional PEPs	Half-life	MBS radius	MBS vol./#p	BE (MeV)
Neon-23	+3	37.140 s	4.72159	19.1702	182.97
Sodium-23	+1/+1 Req.	Stable	4.15764	13.6838	186.56

This is a normal case.

Isotope	Additional PEPs	Half-life	MBS radius	MBS vol./#p	BE (MeV)
Sodium-24	+2/+1 Req.	14.96 h	4.71727	18.3210	193.52
Magnesium-24	0	Stable	4.15616	12.5301	198.26

This is a normal case.

Isotope	Additional PEPs	Half-life	MBS radius	MBS vol./#p	BE (MeV)
Magnesium-27	+3	9.435 min	4.91315	18.3995	223.12
Aluminum-27	+1/+1 Req.	Stable	4.77620	16.9034	224.95

Normal case with little gain in binding energy.

Isotope	Additional PEPs	Half-life	MBS radius	MBS vol./#p	BE (MeV)
Aluminum-28	+2/+1 Req.	2.245 min	4.83036	16.8605	232.68
Silicon-28	0	Stable	4.97732	18.4466	236.54

This is a normal case.

Isotope	Additional PEPs	Half-life	MBS radius	MBS vol./#p	BE (MeV)
Silicon-31	+3/+1 Req.	2.62 h	5.38291	21.0755	262.21
Phosphorus-31	+1	Stable	5.06925	17.6018	262.92

Normal case with little gain in binding energy.

Isotope	Additional PEPs	Half-life	MBS radius	MBS vol./#p	BE (MeV)
Phosphorus-32	+2	14.28 d	5.37951	20.3782	270.85
Sulfur-32	+1 Req.	Stable	5.36431	20.2061	271.78

Normal case with little gain in binding energy.

Isotope	Additional PEPs	Half-life	MBS radius	MBS vol./#p	BE (MeV)
Sulfur-35	+2/+1 PEP gap/ +1 Req.	87.37 d	5.51502	20.0752	298.83
Chlorine-35	+1/+1 Req.	Stable	5.36431	18.4741	298.21

Conflicting case with the binding energy going down slightly while the resulting structure is more densely packed.

Isotope	Additional PEPs	Half-life	MBS radius	MBS vol./#p	BE (MeV)
Chlorine-36	+2/+1 Req.	3.013×10^5 y	5.37909	18.1097	306.79
Argon-36	0	Stable	5.51502	19.5176	306.72

A very strange case: less densely packed as well as a slight downgrade in binding energy. Consequently, the source half-life is high.

Isotope	Additional PEPs	Half-life	MBS radius	MBS vol./#p	BE (MeV)
Sulfur-37	+4/+1 PEP gap/ +1 Req.	5.05 min	6.06738	25.2866	313.018
Chlorine-37	+3/+1 Req.	Stable	5.73346	21.3372	317.100

This is a normal case.

Isotope	Additional PEPs	Half-life	MBS radius	MBS vol./#p	BE (MeV)
Chlorine-38	+4/+1 Req.	37.24 min	5.73346	20.7757	323.208
Argon-38	+2	Stable	5.51502	18.4904	327.343

This is a normal case.

Isotope	Additional PEPs	Half-life	MBS radius	MBS vol./#p	BE (MeV)
Argon-39	+3	269 y	5.62180	19.0831	333.941
Potassium-39	+1	Stable	5.52028	18.0679	333.724

A conflicting case—the resulting structure is more densely packed but the binding energy goes down slightly.

Isotope	Additional PEPs	Half-life	MBS radius	MBS vol./#p	BE (MeV)
Potassium-40	+2	1.248×10^9 y	5.82057	20.6502	341.524
Calcium-40	0	Stable	5.49786	17.4025	342.052

Strange case, the resulting structure is more densely packed and the binding energy goes up slightly, but the half-life is high.

2.17.2 β+ decay

Now we will have a look at all the steps involved in β+ decay, moving from one element to the previous element, at least for those we covered on our journey of the elements up to calcium. Such steps are much rarer—only the step from potassium-40 to argon-40 is of this type in the SAM element buildup scheme to this point.

Isotope	Additional PEPs	Half-life	MBS radius	MBS vol./#p	BE (MeV)
Fluorine-18	0	109.739 min	4.02432	15.1668	137.370
Oxygen-18	+2	Stable	4.06499	15.6313	139.513

Conflicting case, the resulting structure is less densely packed but has a slightly higher binding energy.

Isotope	Additional PEPs	Half-life	MBS radius	MBS vol./#p	BE (MeV)
Sodium-22	0	2.602 y	4.15764	13.6838	174,145
Neon-22	+2	Stable	4.07698	12.9027	177.770

This is a normal case, but with a high source half-life.

Isotope	Additional PEPs	Half-life	MBS radius	MBS vol./#p	BE (MeV)
Chlorine-36	+2/+1 Req.	3.013×10^5 y	5.37909	18.1097	306.789
Sulfur-36	+3/+1 PEP gap/ +1 Req.	Stable	6.06738	25.9890	308.714

Conflicting case: a gain in binding energy, but a much larger structure.

Isotope	Additional PEPs	Half-life	MBS radius	MBS vol./#p	BE (MeV)
Potassium-40	+2	1.248×10^9 y	5.82057	20.6502	341.524
Argon-40	+4	Stable	5.62483	18.6362	343.810

This is a normal case, but with a high source half-life. Potassium-40 to argon-40 requires a β+ decay and emits a gamma ray with 1.460 MeV energy in most cases. This is entirely possible by looking at the difference in binding energy between these two elements. We will investigate the energy balance of these processes in more detail in Sections 7.3 and 11.4.

2.18 ROADS NOT TAKEN

In the process of developing the structure of the elements there were several options available and choices therefore had to be made. We have not yet explained sufficiently why key decisions were made. The short answer is that decisions were made based on the known properties of the elements and their similarities. Let us consider these decisions more deeply.

First, we had to consider the scenario after nitrogen, that is, where to place the next deuteron to arrive first at oxygen and later at neon? We decided to put the four-endings on one side, effectively creating a top side and a bottom side of the growing nucleus. Why not put one four-ending facing up and the other rotated 180° facing down? This idea was given great consideration, but it was realized that it would have opened a second gap at the bottom of the icosahedron, further degrading its stability and creating complications later.

In case of oxygen-17 there were two basic options to position the next PEP. On the V-shaped structure or on the other side. The difference in binding energy gave us a clear indication on the order of positions in which the PEPs had to be placed. We will have a detailed look at binding energy in Sections 7.3 and 9.4.

In the case of fluorine-18 it would have been theoretically possible to create it with one lithium nuclet (Fig. 2.47) to the side (12 + 6 = 18 protons), but this does not play well with a single lithium nuclet being the indicator for an alkali metal— which fluorine is obviously not. The same argument is true for other cases, where the known element properties simply would not fit possible structure options.

Figure 2.47 Fictional fluorine-18 with a carbon nuclet and a lithium nuclet.

As a consequence they were discarded. Another issue that had to be considered was having several possible growth points, with a number of options always being available going forward. Take potassium-39 and calcium-40 as an example. There are three five-endings on argon-38. Which one of those is first converted to a lithium nuclet? Which is second? The path forward therefore became a balancing act between several key factors: spherically dense packing, balancing the nucleus itself, preferred states, and what we actually know about the elements (e.g., valence—see Section 2.19 or the precise emission spectrum of decay steps). This is an ongoing process for the SAM team, a process that has already been through many iterations. Even now we end up with several isomeric configurations for an isotope and have to choose. With more available knowledge or better technology, we may have to revise some of those decisions in the future.

2.19 VALENCE AND OXIDATION STATES

One of the most important experimental findings in chemistry is the so-called valence or oxidation state of an atom. It represents the number of connection points a nucleus offers to electrons of another atom. It can be understood as the capacity of an atom or a group of atoms to connect to other atoms by adding, losing, or sharing outer electrons.

The concept of valence was developed in the second half of the 19th century. Most 19th-century chemists defined the valence of an element as its number of bonds, without distinguishing different types of valence or bond. In 1893 Alfred Werner described transition metal coordination complexes which he distinguished in *principal* and *subsidiary* valences, corresponding to the modern concepts of oxidation state and coordination number, respectively. Originally the oxidation state described the degree of oxidation (loss of electrons) of an atom in a chemical compound. Much later, it was realized that a substance, upon being oxidized, loses electrons, and the meaning was extended to include other reactions in which electrons are lost, regardless of whether oxygen was involved. In 1916, Gilbert N. Lewis explained valence and chemical bonding in terms of a tendency of (main group) atoms to achieve a stable octet of eight valence–shell electrons. According to Lewis, covalent bonding leads to octets by the sharing of electrons, and ionic bonding leads to octets by the transfer of electrons from one atom to another. The term covalence is attributed to Irving Langmuir, who stated in 1919 that "the number of pairs of electrons which any given atom shares with the adjacent atoms is called the *covalence* of that atom." The prefix *co-* means "together," so a covalent bond means atoms share a valence. It is now more common to speak of *covalent bonds* rather than *valence* [Wikipedia 2021/Valence_(chemistry)].

Because this is an experimentally established property of the atom it is of utmost importance for SAM to be in sync with it. As stated before, in SAM the available positive spots of a nucleus, as well as outer electrons, determine the number of connection points an atom carries. In SAM a chemical bond is a bond between an outer electron and a remaining positive spot on the nucleus of another atom. The number of connections that can be made this way accounts for the valence number. Initially we will look at ionic bonds.

We start with a look at the end of the 2nd row of the PTE, where we find neon. We will work our way down from the capping phase from here. Let us consider the 20 triangular faces of the base icosahedron as the positions of the electrons. In neon-20 the 10 inner electrons bound by deuterons and the 10 outer electrons cover these 20 faces completely. It therefore has no positive connection points open, the valence number of neon is therefore 0—there is no connectivity.

The number of positive spots on the nucleus can be quickly identified by considering the number of protons, electrons (total), and triangular faces on the carbon nuclet.

With fluorine there are 9 inner electrons bound to deuterons and 9 outer electrons—covering 18 of the 20 icosahedron faces. Two uncovered faces represent one open connection point. Fluorine therefore has a valence number of –1.

Oxygen has 8 inner electrons bound to deuterons and 8 outer electrons. Therefore, it has 4 uncovered icosahedron faces, giving it a valence number of –2.

Nitrogen has 7 inner electrons bound to deuterons and 7 outer electrons, resulting in 6 uncovered tetrahedron faces and a valence number of –3.

Carbon has 6 inner electrons bound to deuterons as well as 6 outer electrons. Therefore, 8 of the 20 icosahedron faces are uncovered, resulting in a valence number of –4. However, carbon is also the end of the building phase that starts with lithium. Now we consider the beginning of the row but go back one step further. Helium-4 is the element before lithium, represented by a tetrahedron with four faces, all of which are covered by the two inner and outer helium electrons, giving it a valence number of 0.

Lithium has 3 inner electrons bound to deuterons and 3 outer electrons, one more deuteron than the helium-4. This results in a valence number of +1. The same thing is true for sodium, it has one more deuteron than neon, and therefore again a valence of +1.

The scheme is now clear, beryllium has a valence number of +2, boron +3, and carbon—in addition to –4—has a valence number of +4 if moving from lithium upwards. This is the famous so-called cycle-of-eight. This logic is directly seen in the structures. Also remember that beryllium can be viewed as two lithiums intertwined, boron can be viewed as three lithiums intertwined, and carbon can be viewed as four lithiums intertwined (Section 2.12.4).

To summarize, in order to have a chemical connection, not only must we have electrons to share between two atoms, but also each one must have a connection point to the nucleus available—they cannot be noble. The number of positive spots on the nucleus are always to be found on incomplete endings, that is, the lithium nuclet, beryllium ending, boron ending, carbon nuclet (building phase), and for the capping phase with one to four deuterons on the sides of the carbon nuclet.

The negative and the positive attributes always identified in chemistry (ionic bonds) can therefore be interpreted as follows: the minus represents the capping phase—each deuteron in the capping phase that is not yet present represents a valence of –1. Each deuteron added to a noble gas equates to a +1 valence.

When we look at heavier elements, each typically recognizable ending on an active carbon nuclet, like the lithium nuclet, carries that particular valence value, as does a carbon nuclet or the oxygen configuration. This means that in a larger element such as magnesium, which has two lithium nuclets, we can instantly recognize that this atom should (and does) have a valence of +2. A silicon that has a carbon nuclet just like carbon itself carries a valence value like carbon. Figure 2.48 highlights the oxidation states for the elements of the first three rows of the PTE.

Element			Negative states					0	Positive states									Group	Notes
			-5	-4	-3	-2	-1	0	+1	+2	+3	+4	+5	+6	+7	+8	+9		
Z																			
1	hydrogen	H					-1		+1									1	
2	helium	He																18	
3	lithium	Li							+1									1	[16]
4	beryllium	Be						0	+1	+2								2	[17][18]
5	boron	B	-5				-1	0	+1	+2	+3							13	[19][20][21]
6	carbon	C		-4	-3	-2	-1	0	+1	+2	+3	+4						14	
7	nitrogen	N			-3	-2	-1		+1	+2	+3	+4	+5					15	
8	oxygen	O				-2	-1	0	+1	+2								16	
9	fluorine	F					-1											17	
10	neon	Ne																18	
11	sodium	Na					-1		+1									1	[16]
12	magnesium	Mg							+1	+2								2	[22]
13	aluminium	Al				-2	-1		+1	+2	+3							13	[23][24][25]
14	silicon	Si		-4	-3	-2	-1	0	+1	+2	+3	+4						14	[26]
15	phosphorus	P			-3	-2	-1	0	+1	+2	+3	+4	+5					15	[27]
16	sulfur	S				-2	-1	0	+1	+2	+3	+4	+5	+6				16	
17	chlorine	Cl					-1		+1	+2	+3	+4	+5	+6	+7			17	[28]
18	argon	Ar						0										18	[29]

Figure 2.48 The valence and oxidation states of elements up to argon.

[Wikipedia 2021/Oxidation_state#List_of_oxidation_states_of_the_elements]

The numbers we see for the first cycle-of-eight show that the +2 value for magnesium can be +1 and +2. The +2 is marked dominant and is expected as it correlates with the position in the PTE, yet the +1 needs explanation too.

The magnesium nucleus has two lithium nuclets as mentioned before. We think these two nuclets can make a connection between themselves as well as with nuclets from other nuclei (normal chemistry). This results in a singular connection between the two nuclets via one outer electron on the same nucleus. That electron is no longer able to share a connection, thus reducing the available connections by one to other nuclei. Lithium does not show a –1 value, but all the others in the same group do, that is, Na, K, Rb, and Cs all have a –1 or +1 value. The +1 value is the default value though.

The reason endings sometimes have diverging valence values is not always clear. It seems that switching the sign, which represents acceptance or donation of an outer electron to the other atom, is most likely dependent on the combination with other elements. This is known chemistry, and we point in that direction for more research regarding this topic.

2.20 ELEMENT SIMILARITIES AND THE CYCLE-OF-EIGHT

The periodic table of elements shows a regularity that was immediately identified at its inception: the recurring cycle-of-eight. This cycle was called the "law of octaves" by J. A. R. Newlands in 1865. He noticed that when elements are arranged by increasing weight, every eighth element has similar properties.

He suggested this resembled the eight notes of the musical scale, hence its name. This discovery led to further advancements being made in the PTE. Both the 2nd and 3rd row of the periodic table have eight elements.

However, *first* let's ask some big questions: What defines oxygen as oxygen? What defines carbon as carbon? What defines any other element and its properties? These questions are not easily answered and nowhere can a simple and readily available answer to each of them be found.

SAM offers a novel approach to resolving simple questions that require complex answers . . . not difficult, but complex . . . due to the number of possibilities, scaling factors, the dual nature of things, and balance requirements. SAM attempts to find answers to these questions without the need for vast mathematical models. Also, we do not compare mathematical formulae with real, physical things—we are not giving preference to the math when doubt arises. We are convinced the approach to prefer math over reality is wrong and reality is simply what it is. We can use math and language to express ourselves or to describe things, but that's all. First and foremost, we need to stick to the "trivium–quadrivium" method.

Starting with observations from the PTE or rather the properties of the elements, we can conclude the following:

- Columns show similar properties.
- Rows show (initially) the cycle-of-eight.

- Each element has on average two more protons than the previous element—bound by an additional inner electron.
- We also note that the regularities we observe never hold true for the entire PTE—the pattern breaks at some point.

When one takes a closer look at the periodic table of elements, one observes many regularities. The columns in the PTE show elements with similar properties like the noble gases which are all chemically inert. We discussed earlier (Section 2.9) how nobility is expressed through the structure of the nucleus and its relation to the outer electrons.

An even closer look however reveals that elements in the columns are similar in many cases, but not all. The same goes for the cycle-of-eight—working beautifully in the 2nd row, a bit less perfectly in the 3rd row, and by the time we reach the 4th row of the PTE at potassium we see that it seemingly disappears. Looking at the structure of the nucleus using SAM we can see that from sulfur onward, there are now three positions or places where the nucleus can grow further by creating more branches. All three potential branches represent an opportunity for the initiation of an entire cycle-of-eight. What we can conclude is that the 4th row in the periodic table of elements is therefore a combination of the growth of three cycles-of-eight taking place at the same time, creating the observed long row and the transition metals which have properties that appear not to correspond to the cycle-of-eight.

SAM is very helpful in understanding how the properties of the elements come into being. The beginning of the PTE clearly shows the cycle-of-eight. The left side of the periodic table of elements represents the buildup phase of the cycle. All these elements are in the buildup phase of a carbon nuclet. The right side of the PTE represents the noble gases which are complete and inert. Directly left of this column we have the halogens which are all in the capping phase and almost complete.

On the left side of the PTE we can see positive elements and the right side represents negative elements, according to their valence. The reason for this is that elements in the capping phase are close to completion or "perfection" and need, in the case of the halogens, only one more outer electron to be in a noble state—mimicking the noble gases. Chlorine needs one more electron to be like argon. Sodium on the other hand has the base of neon plus one more deuteron in its nucleus and therefore one more outer electron. This outer electron can be shared with the chlorine making the latter complete in outer electrons (wrapped around the carbon nuclet).

Now we see why NaCl (salt) is more noble than its two constituents. The electron is shared by both nuclei and this keeps them together, they bond. An important conclusion here is that to have a chemical bond there must be not only outer electrons that can be shared but electrons that are able to connect to the other nucleus. This can only be done if the nucleus is able to receive a connection, meaning there must be a positive spot left on it to connect to. The moment there are no more positive spots the element is classified as noble—its potential has been taken away through the chemical bond, rendering the remaining atom or molecule inert.

How does the cycle-of-eight relate to valence numbers and the oxidation state? The larger elements, particularly those halfway through the PTE in the center, show great diversity in oxidation states. First, the number and combination of endings is important and is essentially a mixture of these values by adding them up or subtracting them from each other. Then the potential of connecting among each other in the same nucleus causes even more "exceptions" to the logic of the cycle-of-eight. Therefore, there is still indeed a clear recognizable pattern, but the increasing number of growth-points increases the complexity of the nucleus as well as the number and combinations of endings. As we move through the PTE, this causes the diverse attributes and properties of the elements. The strange yet observable pattern here is again a telltale sign for the "mixing up" of simultaneous cycles-of-eight in action.

2.20.1 Oxygen

We come back to the question: What makes oxygen, oxygen? Looking at oxygen, we see that the nucleus is halfway through its capping phase toward neon. With carbon, we see there are four deuterons needed to arrive at neon, with oxygen being halfway with two deuterons placed and two missing. Each deuteron represents one inner electron and one outer electron. Oxygen-16 therefore has eight inner electrons and eight outer electrons—the number of missing inner electrons to reach the noble neon state is 2. There are four positive spots or connection points left (eight in carbon) as each outer electron is connected to two points on the nucleus.

> When we encounter oxygen in nature, it is very often in the form of O_2. Why is that? When two oxygen atoms bond as described above, the connection is made both ways, every positive spot, every face is covered. O_2 therefore resembles the noble state of neon. Oxygen connects to another oxygen to reach a preferred, more noble state. Through this it becomes more or less inert, has no available connection points, cannot create bigger structures—it is therefore a gas.

The reduced number of connection points through this form of molecule creation is why close to the column of the noble elements the melting point goes down. We see gases, liquids, or not very rigid metals, whereas in the middle of the row we have very rigid elements with a lot of possible connections and high valence numbers. With multiple connection points they can never be noble, they interact with a lot of different compounds in chemical bonds.

We therefore can see, understand, and explain why oxygen is what it is, with the properties and attributes it has. This becomes evident by knowing its structure and constituent components. Moreover, we understand why oxygen is so different (gaseous) from the other elements in the

Figure 2.49 Argon-36 with nuclet-bounding spheres.

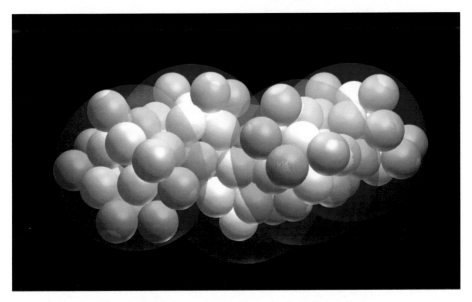

Figure 2.50 Krypton-80 with nuclet-bounding spheres.

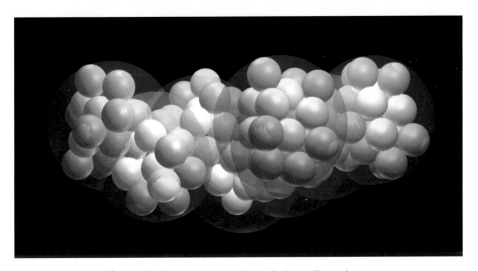

Figure 2.51 Xenon-128 with nuclet-bounding spheres.

"oxygen group," so different that it is often left out when characterizing the group. As an element of the 2nd row it is close to being noble, from the 3rd row forward that is no longer possible.

The noble state of elements can also be visualized with bounding spheres around the carbon nuclets (Figs. 2.49–2.51).

The bounding spheres show a densely packed state.

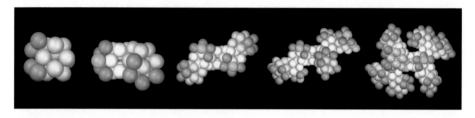

Figure 2.52   **Figure 2.53**   **Figure 2.54**   **Figure 2.55**   **Figure 2.56**
Oxygen-16.   Sulfur-32.   Selenium-74.   Tellurium-122.   Polonium-208.

2.20.2 Oxygen group

This group consists of the elements oxygen, sulfur, selenium, tellurium, and the radioactive element polonium. We discussed oxygen above already. What we can see
from the structure is that the elements in this group all show at least one ending with
a carbon nuclet capped with a four-ending on one side only, similar to oxygen-16
(Figs. 2.52–2.56).

The oxygen group is somewhat strange in the sense that the elements belonging to
this group have very diverging properties. Oxygen is most important for life and in
biology. Sulfur is too, but in a much lower percentage. Selenium is still essential for
life, but only in very low abundances needed for enzymes. Exposure to higher than
needed amounts makes selenium toxic. The other heavier elements in this group, tellurium and polonium, are considered simply toxic and polonium is an alpha emitter
which makes it, once absorbed, very deadly.

Abundance-wise the group shows the very high abundance of oxygen in the universe
and even higher here on Earth, making it the most abundant element for us. Sulfur is
still quite abundant but makes up only 0.04% of the crust of the Earth. Selenium
is even lower in abundance at $5 \times 10^{-6}\%$, tellurium at $1 \times 10^{-7}\%$; and polonium is at
virtually 0% (a trace element).

As already mentioned, the chemical properties of these elements start to diverge
from those of oxygen. Sulfur clearly is not a gas, but it is quite reactive. Selenium is
still considered to be a non-metal, but tellurium is in the metalloid group and polonium
is considered a post-transition metal similar to lead or tin.

The higher the period in the PTE the fewer non-metals we see and the elements
below oxygen start to show more and more metal-like properties even though they still
have some similarities. This shows a toxic effect in biochemistry when organisms are
exposed to the heavier elements in this group and the lighter elements are replaced by
the heavier ones.

2.20.3 Halogens

Halogens are the elements in the PTE in the last column before the noble gases—which
we have already discussed (Section 2.9). They share a similar property of having a –1
value for the oxidation state, and they also tend to bind with themselves in a binary
state (i.e., F_2, Cl_2, Br_2, and I_2) (Fig. 2.57).

halogen	molecule	structure	model	d(X–X) / pm (gas phase)	d(X–X) / pm (solid phase)
fluorine	F_2	F–F 143 pm		143	149
chlorine	Cl_2	Cl–Cl 199 pm		199	198
bromine	Br_2	Br–Br 228 pm		228	227
iodine	I_2	I–I 266 pm		266	272

Figure 2.57 Visual description of the binary state of halogens.
[Wikipedia 2021/Halogen]

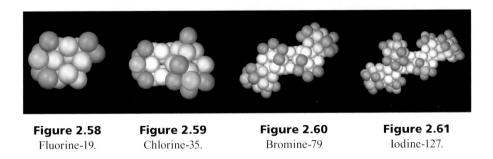

Figure 2.58
Fluorine-19.

Figure 2.59
Chlorine-35.

Figure 2.60
Bromine-79

Figure 2.61
Iodine-127.

Comparing elements in the Atom-Viewer (Appendix G) we see that they are all alike and share the commonality of lacking one capping deuteron with all the other branches being in the noble full capping state (Figs. 2.58–2.61).

If this type of element is part of a chemical bond then the possibility of more elements bonding is significantly limited. This means they cannot make long-chain or ring-like chemical structures. In order to do that we need elements belonging to the carbon group and metals with multiple connection points due to them having multiple active endings.

2.20.4 Carbon group

The carbon group is perhaps one of the best known groups. It contains the elements carbon, silicon, germanium, tin, and lead (Figs. 2.62–2.66).

When comparing the structures of these elements the reason they are so similar becomes clear. We notice, of course, when the elements become heavier that the

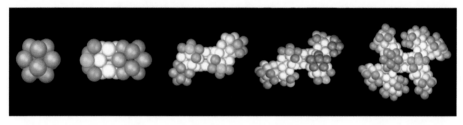

Figure 2.62 **Figure 2.63**   **Figure 2.64**   **Figure 2.65**   **Figure 2.66**
Carbon-12.   Silicon-28.   Germanium-70.   Tin-114.   Lead-204.

complexity of the cycle-of-eight interaction increases. Looking at the structure, tin appears to be quite different. Although the total valence may be similar, it is built up of different ending structures. This seems to be typical for the center of the PTE, that is, the more in the center we find an element the higher the number of combination options, hence a more complex structure appears. Obviously, this increases as we go further down the PTE as the elements grow larger. Lead, the last element in this group, shows a carbon nuclet and an oxygen-type ending. Based on the structure we must also consider that tin is not correctly placed in the group or that the structure of tin is not correctly determined in SAM. This will have to be investigated further.

2.20.5 Metals

This group represents the largest group of the PTE and makes up more than half the known elements, about 65 of them. These elements share more or less similar properties although they vary greatly among themselves. We will next take a closer look at some sub-groups within the metal group.

2.20.5.1 Noble metals

The elements of the noble metals group likely represent the oldest metals known to humankind, since they are the ones that can be found as naturally occurring metals, while almost all other elements are not found in a pure form, especially not the metals. The latter are usually oxidized under normal earthly circumstances or chemically bound to other elements such as fluorine, chlorine, and sulfur. This would be the field of geology.

When comparing the structures of the group of noble metals, it is obvious that they share a similar structure.

They are all composed of incomplete noble endings—not unlike carbon, nitrogen, oxygen, and fluorine. In a way these metals seem to be more similar to elements mentioned above than to other metals (Figs. 2.67–2.70).

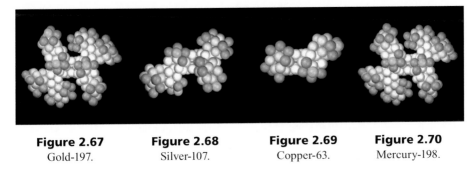

Figure 2.67
Gold-197.

Figure 2.68
Silver-107.

Figure 2.69
Copper-63.

Figure 2.70
Mercury-198.

2.20.5.2 Platinum group

The structures for the nuclei of the elements of the platinum group clearly show correlations again. The main feature is that they are all more or less constructed like oxygen (i.e., with a half-capped carbon nuclet).

That this group is kept separate from the noble metals has historic reasons. The noble attribute comes from the fact that these elements either oxidize very slowly or not at all when exposed to air (Figs. 2.71–2.76).

Gold is not part of the platinum group since it is in its own category, but it is worth mentioning that gold (also silver and copper) and these platinum group elements are often found together in nature—looking at the nuclei we can see why they should be placed in the same group.

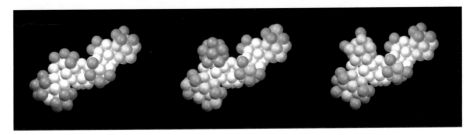

Figure 2.71 Ruthenium-96. **Figure 2.72** Rhodium-103. **Figure 2.73** Palladium-104.

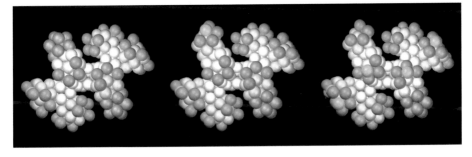

Figure 2.74 Osmium-184. **Figure 2.75** Iridium-191. **Figure 2.76** Platinum-194.

Elements usually seem to have other elements associated with them, as is the case with certain elements in the same group. For example, the elements osmium, iridium, and ruthenium from the platinum group are mostly found in nickel-bearing ores.

2.20.5.3 Alkali metals

These metals are all members of the first column in the PTE and share the property of having an oxidation state of +1. They tend to be highly reactive and low in density. They are also the elements with the largest atomic radii per cycle-of-eight. The common feature of this group is the single lithium nuclet (Figs. 2.77–2.81).

2.20.5.4 Alkaline-earth metals

The alkaline-earth metals are all in the second column of the PTE and have an oxidation state of +2. They are called the alkaline-earth metals because they tend to be the elements that are most abundant in the Earth's crust—with the exception of beryllium (Figs. 2.82–2.86). These minerals tend to be insoluble in water and were therefore called earth metals (oxides).

The structural commonality is the double lithium nuclet—as stated before with beryllium being an exception—but it can be viewed as two lithiums intertwined.

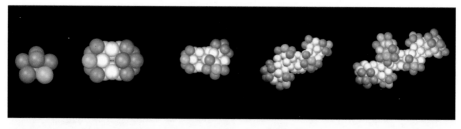

Figure 2.77	**Figure 2.78**	**Figure 2.79**	**Figure 2.80**	**Figure 2.81**
Lithium-7.	Sodium-23.	Potassium-39.	Rubidium-87.	Cesium-133.

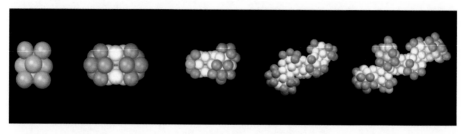

Figure 2.82	**Figure 2.83**	**Figure 2.84**	**Figure 2.85**	**Figure 2.86**
Beryllium-9.	Magnesium-24.	Calcium-40.	Strontium-84.	Barium-130.

2.20.5.5 Transition metals

This group is the largest sub-group of the metals and occupies the entire central region of the PTE: iron, titanium, tungsten, zinc, and copper all belong to this group (Figs. 2.87–2.93). The noble metals mentioned above are part of this group of transition metals and form a sub-group of their own within this group.

Commercially pure (99.2% pure) grades of titanium have an ultimate tensile strength of about 434 MPa (63,000 psi), equal to that of common, low-grade steel alloys, but are less dense. Looking at the structure, we can see that titanium is in effect a combination of iron and a carbon. Steel is iron plus some carbon mixed into it, giving it much greater strength. Titanium seems to be that on its own, but with many fewer protons, giving titanium its unique properties: strong, durable, and lightweight.

The most important thing to note here is that we can see in the structure why this group of transition metals as a whole has so many diverging properties. As we have shown before, we see the effect of multiple branches "growing" at the same time. Several cycles-of-eight are in a different phase of completion on one nucleus. This mixing and combining of endings in different stages is the cause for sometimes seemingly conflicting attributes, as well as being an explanation why this group does not show well-defined similar properties. This group is clearly the most complex group from a structural point of view, due to combination possibilities.

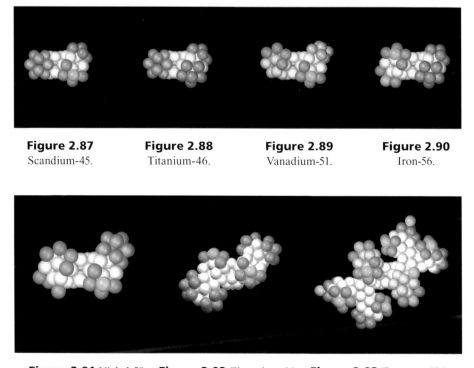

Figure 2.87
Scandium-45.

Figure 2.88
Titanium-46.

Figure 2.89
Vanadium-51.

Figure 2.90
Iron-56.

Figure 2.91 Nickel-58.   **Figure 2.92** Zirconium-90.   **Figure 2.93** Tungsten-184.

2.21 SUMMARY

We will now summarize what we have learned in this chapter while creating and studying the elements:

- The rules of spherically dense packing: each new sphere must be part of a tetrahedron structure. The icosahedron is the ideal structure.
- A nucleus can grow by adding a proton, a proton–electron pair (PEP), or a whole deuteron, or several of these.
- Energetically low states are preferred (higher binding energy).
- The minimum energy state represents the densest (possible) packing.
- Growth of the nucleus happens with a carbon nuclet structure as a backbone.
- The active endings of the backbone show a repeating pattern of sub-structures, going through a building and a capping phase.
- The nucleus must be balanced (structurally as well as in terms of charge).
- Valence and oxidation state are based on the structural configuration of the nucleus.
- Groups of the PTE can be identified by the nucleus having a similar structure.

CHAPTER 3

Heavier elements of the PTE

3.1 BRANCHING AND ELONGATION

When we try to move on with our previous scheme of stepping from nucleus to nucleus, we notice a breakdown. Much still works, but overall the stable isotopes per element are more dispersed, which does not allow us to continue our prior scheme as easily as before. The 18 elements of the 4th row of the standard periodic table of elements (PTE)—potassium through krypton—are more complex, because as the nucleus branches it creates more possibilities in terms of where the nucleus can grow. This is the reason there are many transition metals. In general, during growth the building phase seems to be preferred over the capping phase. This means we see growth at several places on the nucleus in parallel. To get a better understanding about what is happening, we must look at the basic nuclet structure.

In Figure 3.1, after carbon (*blue*), we start from left to right with the slightly distorted core (*black*) carbon nuclet (see Section 2.6), neon is the last element with one nuclet. We then move to sodium which has two nuclets—the even further distorted carbon nuclet and a lithium nuclet (*red*). With sulfur, chlorine, and argon we reach the state of two distorted carbon nuclets.

Potassium, as the next alkali metal, shows again the next lithium nuclet. It is immediately followed by calcium which shows four nuclets (two lithiums) and the first branch in the structure. Copper shows four carbon nuclets, but they are not completely capped—making it close to being noble (Fig. 3.2).

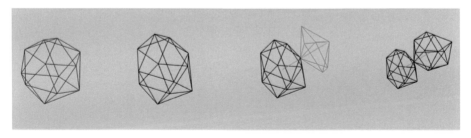

Figure 3.1 Core structure of carbon, nitrogen to neon, sodium, and sulfur to argon.

The last element of the 4th row, krypton, is inert because it is the completion (complete capping) of five carbon nuclets. The branching continues—overall the nucleus elongates in the process of adding new carbon nuclets to the backbone structure. If it did not elongate, we would see colliding branches very early on, and therefore would be unable to create heavier nuclei that exist in nature. If we look at the structure of silver and compare it with the structure of xenon, it becomes clear why silver is considered a noble metal (Fig. 3.3).

Lanthanum is a rare-earth element. It does not yet show the middle branches. Gold too has this very noble state-like backbone structure (Fig. 3.4).

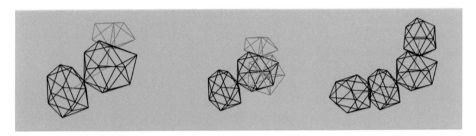

Figure 3.2 Core structure of potassium, calcium, and copper.

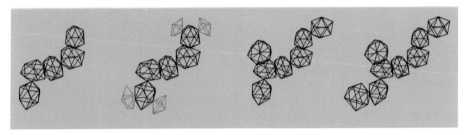

Figure 3.3 Core structure of krypton, zirconium, silver, and xenon.

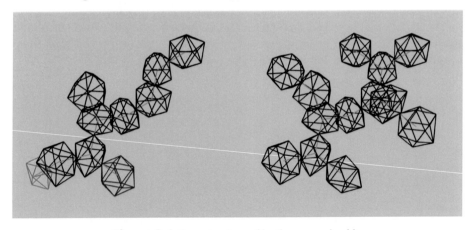

Figure 3.4 Core structure of lanthanum and gold.

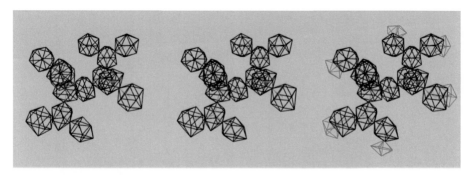

Figure 3.5 Core structure of lead, radon, and uranium.

We stated before that growth appears to prefer the building phase over the capping phase. This explains why most of the elements are metals and why a diagonal line crosses the chemical periodic table of elements between the transition metals and non-metals. With gold, the structure of the nucleus already looks very complicated, as if branches will collide if the nucleus grows bigger.

3.2 COLLIDING BRANCHES

Collision, or close contact, of branches happens just after lead, which "coincidentally" is one of the last stable elements (Fig. 3.5, *left*).

Radon is the next noble element but is unstable. Uranium is the last element for which we show the structure here. The problem with two branches coming too close to each other is the main argument against returning element stability (Fig. 3.5), which is an expectation of the Standard Model.

We can therefore conclude that there is no island of stability after lead.

3.3 LOOKING AT LANTHANIDES AND ACTINIDES

The 4th and 5th row of the PTE presented us with a lot of complexity when we realized three cycles-of-eight were running in parallel. It gets worse with the 6th and 7th row: here we have *six* cycles-of-eight running in parallel in each row.

3.3.1 Lanthanides

The lanthanide series of chemical elements comprises the 15 metallic chemical elements from lanthanum through lutetium. These elements, along with the chemically similar elements scandium and yttrium, are often collectively known as the rare-earth elements. The "earth" in the name "rare-earth" arises from the minerals from which they were

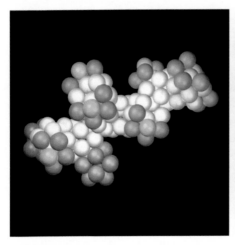

Figure 3.6 Structure of a lanthanum-139 nucleus.

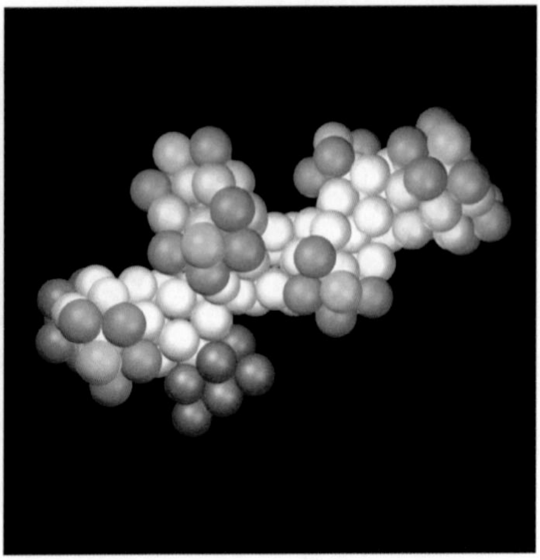

Figure 3.7 Structure of a cerium-140 nucleus.

isolated, which were uncommon oxide-type minerals. The "rare" in the "rare-earth" name has much more to do with the difficulty of separating out each of the individual lanthanide elements than with the scarcity of any of them. Cerium is for example the 26th most abundant element in the Earth's crust and more abundant than copper. All the lanthanide elements are commonly known to have +3 and +2 oxidation states.

Lanthanum, after which the series is named, is shown here with 139 protons (Fig. 3.6). The nucleus is elongated while the branches are not very big, making it one of the least dense packed nuclei when comparing the volumes of minimal bounding spheres (MBS).

The next element in the series is cerium, here with 140 protons (Fig. 3.7). We can see that the two middle branches to the right have just started to grow. Those middle branches will later contribute to forming the first two collision points.

3.3.2 Actinides

Just before the actinides we find francium and radium. Like their other cycle-of-eight counterparts, they show a valence of +1 and +2 respectively, and they fall in those groups (alkali metals and alkaline-earth metals).

All actinides are radioactive and release energy upon decay; naturally occurring uranium and thorium, and synthetically produced plutonium are the most abundant actinides on Earth. Uranium and thorium have diverse current or historical uses, and americium is used in the ionization chambers of most modern smoke detectors. The reason they are called the actinides is because of their similar attributes to actinium, although this is more of an historic understanding. Looking at the nuclear structure in SAM we see that they are in effect continuing to grow by adding lithium nuclets. The oxidation state of actinium is +3, thorium +4, protactinium +5, uranium +6, and neptunium +7.

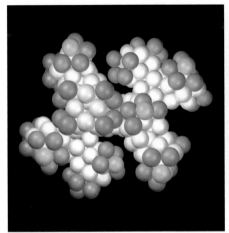 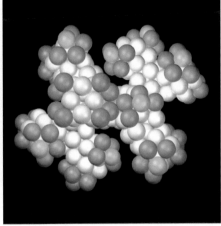

Figure 3.8 Structure of an actinium-227 nucleus.

Figure 3.9 Structure of a thorium-232 nucleus.

This incremental growth keeps occurring until a nucleus is saturated with lithium nuclets in all directions and the nucleus can handle no more, before collapsing in on itself. This is the point where the normal, naturally occurring elements have their boundary and no additional heavier elements are possible. The nucleus is "collapsing in on itself" because the space where the next carbon nuclet should go is already occupied by another branch. The culprit is the 3rd collision point that is formed around lead. The collapse leads to nuclear decay (α particle release), fission, and spallation.

Actinium is the first of the actinides, here shown with 227 protons (Fig. 3.8). The growth of the middle branches is now clearly visible since they are fully developed.

Thorium-232 is the second actinide (Fig. 3.9). It is well known for the thorium cycle, which we will discuss in Section 13.6.

3.4 WHAT HAPPENS WITH HEAVIER ISOTOPES?

One important thing that happens is the growth of the two middle branches. At the start of the actinides group (element actinium) growth on those branches is mostly complete. Not much more is possible, otherwise the branches would touch or—worse—overlap. This is the reason for the natural radioactivity of the actinides: the structure has grown as much as possible in those places and two branches in the center are too close together. Even the slightest exposure to external energy will make the nucleus susceptible to breaking apart.

We conclude that the elements above lead are unstable because the nucleus can no longer achieve a stable configuration. The reason there are still elements above lead is that the nucleus is more or less "overloaded" in a semi-stable state. It can do this since the outer endings can structurally still hold the next viable configuration. This is shown in the actinides that grow lithium nuclets step by step. It does however mean that the threshold of stability at the larger level is at odds with the structure. The nucleus becomes elongated because if it does not it will decay much sooner. However, there is a limit.

> The last unstable, but still structurally sound, element that can be created with SAM is americium. Elements above americium are artificial and can only be created by fusing nuclei together with brute force. The resulting nucleus is without structure that can be described with SAM and only exists for a few milliseconds before fissioning again.

3.5 THE RULES OF THE BACKBONE STRUCTURE

It is time to look in more detail at the backbone structure of the nucleus and its theoretical fractal growth pattern. We start with the central carbon-like icosahedron structure with 12 spheres. It is a spherical-symmetrical structure, but once we attach the first deuteron to it, a gap opens that distorts the structure and determines the location of the two new growth points. In fact the whole structure and its fate is determined by this first distortion step.

Those two points can grow to become carbon-like icosahedron structures too, but because of the shared sphere they each provide only 11 spheres. Each of those, once fully grown, provides another two new growth points. Table 3.1 shows the ensuing pattern with the number of icosahedrons, the number of core spheres, and the number of capping and balancing spheres.

The fractal geometry of the backbone is complicated. Those first two connection points on the initial icosahedron show an unusual angle of ~31.5° each (because of the top-gap). The two connecting icosahedrons are tilted and rotated by 36°. The orientation of the two icosahedrons and where they themselves create new connection points ensures maximum separation of the two new branches. This geometric setup repeats with each new icosahedron, but it cannot go on forever as the structure folds in on itself. The fractal growth pattern breaks down just after the creation of the 15th icosahedron. Once all other growth points are used up, element creation comes to its natural end.

We also see a strong indicator for missing elements as there are at least 10(!) missing noble configurations based on the structure alone. This also shows the three and/or six cycles-of-eight running in parallel, as each cycle can reach a noble state. When three cycles-of-eight are running in parallel we see two missing noble configurations, when there are six running we see five missing noble configurations (Table 3.1, last column). If they are missing, what is the reason? Why do we not see those elements in nature? They could be unstable. If, as we assume from observation, the building phase is preferred over the capping phase, then nature would simply skip them. We will look into this in more detail in Section 14.2.

Table 3.1 Building the backbone with icosahedrons with initially 12, then 11 protons.

# Icosahedron	Pattern	# Core spheres	# Capping and balancing spheres
0	0	0	4 + 0 (helium-4)
1	12	12	8 + 0 (neon-20)
2	12 + 11	23	12 + 1 (argon-36)
3	12 + 11 + 11	34	16 + 0 (missing 50)
4	12 + 11 + 11 + 11	45	20 + 1 (missing 66)
5	12 + 11 + 11 + 11 + 11	56	24 + 0 (krypton-80)
6	12 + 11 + 11 + 11 + 11 + 11	67	28 + 3 (missing 98)
7	12 + 11 + 11 + 11 + 11 + 11 + 11	78	32 + 2 (missing 112)
8	12 + 11 + 11 + 11 + 11 + 11 + 11 + 11	89	36 + 3 (xenon-128)
9	12 + 11 + 11 + 11 + 11 + 11 + 11 + 11 + 11	100	40 + 2 (missing 142)
10	12 + 11 + 11 + 11 + 11 + 11 + 11 + 11 + 11 + 11	111	44 + 1 (missing 156)
11	12 + 11 + 11 + 11 + 11 + 11 + 11 + 11 + 11 + 11 + 11	122	48 + 0 (missing 170)
12	12 + 11 + 11 + 11 + 11 + 11 + 11 + 11 + 11 + 11 + 11 + 11	133	52 + 1 (missing 186)
13	12 + 11 + 11 + 11 + 11 + 11 + 11 + 11 + 11 + 11 + 11 + 11 + 11	144	56 + 2 (missing 202)
14	12 + 11 + 11 + 11 + 11 + 11 + 11 + 11 + 11 + 11 + 11 + 11 + 11 + 11	155	60 + 1 (radon-216)
15	12 + 11 + 11 + 11 + 11 + 11 + 11 + 11 + 11 + 11 + 11 + 11 + 11 + 11 + 11	166	64 + 2 (missing 232)

3.6 SUMMARY

We will now summarize what we have learned in this chapter while looking at heavier elements:

- To prevent colliding branches the nucleus elongates, but there is a limit to this. Due to the branching and the densest packing being at odds with each other, the nucleus cannot grow beyond a certain point.
- There is no island of stability after lead.
- There is a strong indication for missing elements, especially missing noble gases.

This leads us directly to the organizational patterns of the nucleus we constructed while growing the nucleus:

1st order = single nuclet. The 1st-order pattern is limited due to the maximum platonic solid (icosahedron/dodecahedron) which is the structure of carbon. This follows from the definition of platonic solids.

2nd order = two nuclets. This is the doubling factor that creates the fractal of the larger elements. Growth must be balanced, otherwise the nucleus will not be stable. The last nucleus with only 2nd-order organization is phosphorus.

3rd order = fractal growth on several growth points. This happens while following the fractal growth pattern which is competing with the spherical densest packing rule. The protons added will follow all the previous rules where possible. This is a balancing act between the branches, related to nuclear reactions, isotope stability, and decay rates. In effect the 3rd-order pattern steers the distribution between the branches while the 2nd-order is active between endings.

4th order = elongation. Elongation as a 4th-order pattern happens to prevent colliding branches as long as possible.

## CHAPTER 4

Further advancements

4.1 LOADING ELEMENTS WITH PROTON–ELECTRON PAIRS

We have discussed carbon before, and we recognized the special role its structure plays in creating the backbone of a nucleus. There are three naturally occurring isotopes of carbon on Earth: carbon-12, which makes up 99% of all carbon; carbon-13, which makes up 1%; and carbon-14, which occurs in trace amounts, making up about 1 or 1.5 atoms per 10^{12} atoms of carbon in the atmosphere. Carbon-12 and carbon-13 are both stable, while carbon-14 is unstable and has a half-life of 5,730 ± 40 years. Carbon-14 decays into nitrogen-14 through β– decay. There exist, of course, other artificial isotopes of carbon created through technology.

We also concluded from the icosahedron structure of carbon-12 that carbon-20 must be the isotope with the maximum number of additional proton–electron pairs (PEPs) to be firmly attached to the nucleus. More does not work since all possible spots are taken. Moreover, this is exactly what typical isotope lists for carbon show—at least if we ignore isotopes that are not purely based on experimental data (which means they are theoretically derived) and those with so-called "halo neutrons," which means they are not really attached to the nucleus. There are also those isotopes that directly emit a "neutron" after a short period of time. They count only as weak connections. This creates a number corridor between where the added "neutrons" are easily removable and where we enter the theoretical domain. In addition to that we have already seen that adding PEPs, beyond the necessity to stabilize the nucleus, quickly starts to destabilize it, resulting in β– decay.

Now the question must be: Can we determine the maximum numbers for each element from SAM? Below carbon-12 it is a bit arbitrary, carbon-12 we already know. Above carbon-12 we can apply our rules of isotope creation based on nuclets and endings (Table 4.1). We identified connection points for PEPs on a nucleus. PEPs added to those points we call strong additions. PEPs added on top of those we call weak additions. The last two columns of Table 4.2—which arbitrarily ends with krypton—show the strong/weak corridor in SAM and the direct emitters/theoretical domain corridor of the Standard Model as described above.

A few example isotopes with the maximum number of PEPs will be shown here, constructed from the base element configuration by applying the rules of Table 4.1. The four-endings of neon-20 take one PEP (making it a five-ending) and then three more are positioned on each side (Fig. 4.1). With silicon-39 we have eight additional

Table 4.1 Maximum number of PEPs that can be added to endings/nuclets.

Nuclet/ending	Strong adds	Weak adds	Remarks
Empty side on a carbon nuclet	4	0	Max. two sides
Deuteron (two-ending)	2	0	
Four-ending	1	3	Will become five-ending + adds
Five-ending	0	3	
Lithium nuclet	5	0	
Beryllium ending	4	0	
Boron ending	4	0	
Carbon	8	0	
Final ending (backbone)	0	0	

Table 4.2 Maximum number of PEPs to be added to elements.

Isotope	Strong adds	Weak adds	SAM strong/weak	Standard Model direct emitters/max non-theoretical isotope
Helium-4	0	0	4/4	4/4
Carbon-12	8	0	20/20	15/20
Nitrogen-14	6	10	20/20	16/21
Oxygen-16	5	3	21/24	21/26
Fluorine-19	2	3	21/24	21/29
Neon-20	2	6	22/28	25/31
Sodium-23	5	3	28/31	26/34
Magnesium-24	10	0	34/34	29/37
Aluminum-27	4	3	31/34	30/38
Silicon-28	8	3	36/39	34/41
Phosphorus-31	12	0	43/43	37/43
Sulfur-32	5	6	37/43	40/45

Isotope	Strong adds	Weak adds	SAM strong/weak	Standard Model direct emitters/max non-theoretical isotope
Chlorine-35	2	6	37/43	42/46
Argon-36	2	9	38/47	46/48
Argon-38	0	9	38/47	46/48
Potassium-39	5	6	44/50	47/53
Calcium-40	10	3	50/53	51/54
Scandium-45	13	3	58/61	52/55
Titanium-46	18	0	64/64	55/57
Vanadium-51	9	6	60/66	55/60
Chromium-50	15	3	65/68	60/62
Manganese-55	6	9	61/70	61/65
Iron-56	10	6	66/72	65/68
Cobalt-59	4	9	63/72	68/71
Nickel-58	10	9	68/77	71/72
Copper-63	6	12	69/81	72/76
Zinc-64	11	9	75/84	78/80
Gallium-59	14	9	73/82	79/81
Germanium-70	14	6	84/90	82/82
Arsenic-75	15	9	90/99	82/82
Selenium-74	10	12	84/96	86/88
Bromine-79	6	15	85/100	86/92
Krypton-80	6	18	86/104	91/93

PEPs on the carbon nuclet and three more on the five-ending (Fig. 4.2). Sulfur has a half-empty carbon nuclet with 4 strong adds as well as one strong addition on the four-ending and 6 weak additions (Fig. 4.3). Argon 47 has the endings of the argon core structure filled (Fig. 4.4). With calcium-53 the lithium nuclets are fully filled with PEPs as well as the five-ending (Fig. 4.5). The same is true for iron-72 (Fig. 4.6). At some point this scheme will break down because not all the theoretically valid points we count above are viable. Branching is, of course, the reason since it will bring

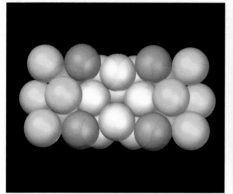

Figure 4.1 Neon-28.

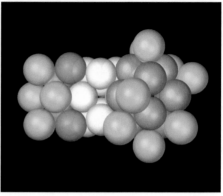

Figure 4.2 Silicon-39.

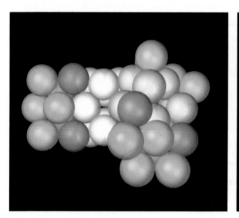

Figure 4.3 Sulfur-43.

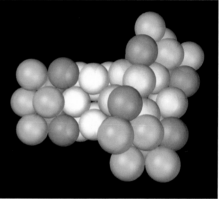

Figure 4.4 Argon-47.

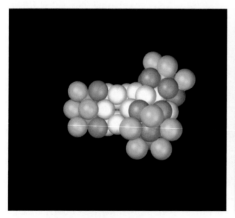

Figure 4.5 Calcium-53.

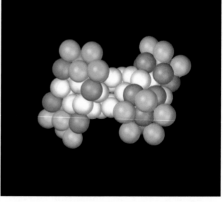

Figure 4.6 Iron-72.

parts of the nucleus, through additional PEPs, too close together until they eventually overlap, something that is not allowed. However, we could easily apply the scheme beyond krypton if necessary.

When comparing the maximum number of "strongly" connected extra PEPs and the ones that can still be added, but only "weakly," to the literature values (Table 4.2) we notice that the maximum as seen in observations and the SAM maximum seem to at least correlate. This provides a good indication that structures defined by SAM are correct.

Obvious questions with so many PEPs are: What is the preferred spot that a PEP would occupy on a nucleus, and is there a logic, and what is it? What is the sequence of adding PEPs to valid spots for a nucleus? There are only partial answers to these questions yet—based on binding energy steps. One observation related to this question is that the first two weak adds, onto four- and five-endings, seem to happen either together or not at all. The answers would be important not only for overloading nuclei with PEPs but also for typical isotopes of elements. The sequence of adding PEPs to a nucleus changes its shape. It determines where the nucleus grows next. If that sequence is not determined, then there are several options for the nucleus to grow. Our initial definition of a nuclear isomer did not exclude structural differences. This is irrelevant for the Standard Model, but it is relevant in SAM (see Section 2.18). We need to refine the definition of the isomer.

4.2 REFINING THE ISOMER DEFINITION

Initially, we defined an isomer as a variation of an isotope having the same components (number of deuterons plus single protons, number of PEPs) but having them arranged in a different way. Taking this into account, we can now discern sub-types of isomeric configurations:

- If the basic nucleus structure (backbone, ending positions, and ending types) is the same, but PEP positions are slightly different, we deal with the excited (metastable) isotope definition of the Standard Model. The excitation of a nucleus has something to do with this sub-type of isomers. We call this a *base structure isomer*.
- If the ending types are still the same, but the positions change in addition to modified PEP positions, we have a different isomer sub-type before us. We will call this an *ending type isomer*.
- If also the ending types change, in addition to ending positions and modified PEP positions, we can identify another isomer sub-type. We will call this a *component isomer*.

Isotopes are in general a difficult topic because there are many possibilities when we look for example at an element such as palladium. A larger or heavier element has multiple PEP connection points. The maximum isotope is about eight PEPs heavier than the first stable isotope. So, we can have any number of extra PEPs ranging from one to eight, and they can occupy multiple locations. Depending on their number and location, this would create a stable or an unstable configuration. The stable ones are usually more or less clearly defined, the unstable ones seem to represent the so-called "excited states" of atomic nuclei (isotopes). Those are what cause an isomeric transition (IT), or as we like to describe it: a resettling of the nucleus by moving around protons driven by the densest packing rules and the attracting/repelling forces of the nucleus.

In this way SAM naturally and inherently illustrates and reflects the complexity that characterizes the field of isotopes. We cannot (yet) look into the nucleus to determine its structure. The current configurations of the nuclei, as presented in this book, are based on the available information, and we are looking for ways to test and possibly falsify such configurations. Precise decay steps may help us, as does looking at binding energy steps. Some advancement of technology perhaps might help at some point in the future too.

4.3 NOTATION SYSTEMS FOR ELEMENTS

So far, we have used the number of protons (or nucleons) as well as the element name to properly name an isotope (e.g., zirconium-88). However, it is common practice to use additional numerical designations for fully identifying a certain element—according to the Standard Model—such as $_{88}Zr^{40}$ or $^{88}Zr_{40}$. The total number of protons cannot be used for this in SAM, as there are many more in the nucleus, as part of PEPs—compared with the Standard Model. Instead, we used the number of deuterons and single protons in the nucleus. This defines the element too (excluding variations in shape), presumably in accordance with the Standard Model. However, we can imagine widely differing structures with an identical number of deuterons, single protons, and PEPs. Those would clearly be different elements based on their structure with different properties. We have already seen candidates—the missing noble gases. In the end, a notation based on the element number will not be enough to clearly define an element, additional structural information is required—currently the name of the element which corresponds to a specific shape. One can now be certain that a count of outer electrons—as in the Standard Model—will not be enough to distinguish elements. However, we will at least consider the deuteron and single proton count or a modification thereof as a sorting mechanism and see where we end up.

4.4 THE DEUTERON AND SINGLE PROTON COUNT

Does the element definition based on the deuteron and single proton count actually work? In Section 2.16 we discussed the impact of a whole carbon nuclet on the deuteron/single proton count. Silicon is the first instance where we have a complete additional carbon nuclet in the nucleus and therefore also a single proton. We have seen single protons before with hydrogen-1 and helium-3, but these are not part of a carbon nuclet as a bigger structure. The atomic number of silicon in the Standard Model is 14. No additional PEP should be there, but clearly we see in Figure 4.7 a five-ending, which includes an additional PEP. What is going on?

When we count the number of protons in silicon-28 (Fig. 4.7) we have $12 + 11 + 2 + 2 + 1 = 28$. That is correct. When we count the deuterons, we only have $6 + 5 + 2 = 13$.

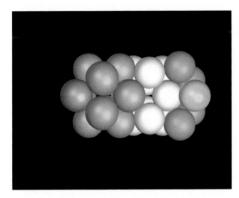

Figure 4.7 Silicon-28 with additional single proton marking (*brown*).

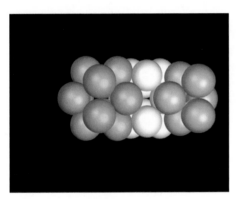

Figure 4.8 Phosphorus-31 with additional single proton marking (*brown*).

Figure 4.9 Phosphorus-31 from the side with one quasi-inner electron and single proton marked in *brown*.

Is one deuteron missing? However, there is a single proton we can count instead. The bluish carbon nuclet carries five deuterons and an additional proton (= 11 protons), which is marked brown in Figure 4.7, although its actual position is unclear. The given position at the point in time of writing this book is an educated guess. The proton would also provide the additional missing outer electron, bringing the count to 14. The five-ending carries an additional PEP which provides an inner electron. So, we actually end up with 14 outer and 14 inner electrons. To keep up with the numbering scheme of the Standard Model we need to count the single protons too. This does not come unexpectedly but is confirmed here.

The next element—phosphorus—presents another issue. With atomic number 15 for phosphorus-31 we should see 15 outer electrons in the Standard Model, which would imply 15 deuterons (15 inner electrons)/single protons and a PEP (Fig. 4.8).

We see $12 + 11 + 8 = 31$ protons. The number of deuterons is $5 + 6 + 4 = 15$. However, there is no PEP, just a proton (brown), seen on the left. This means one more outer electron exists, actually one too many, the count is 16 now. However, we are still missing one inner electron.

We think that because of the setup shown with the proton one outer electron is pulled in very close to the nucleus and becomes a "quasi-inner electron"—in a way providing the inner electron for the proton, to become a semi-PEP (Fig. 4.9). One anchor point of this electron is probably the whole carbon nuclet. The other anchor point is a deuteron on the beryllium ending. Therefore, we have 15 outer electrons and 16 inner electrons (one of them a quasi-inner electron). This is a new class, a new state of electrons, something we do not introduce lightly . . . but stay with us for a moment.

Why did this not work with silicon? On the right side of the nucleus there is only a five-ending. That is not enough structure to support a quasi-inner electron on top of a carbon nuclet. The electron related to the additional carbon in the nucleus has no choice but to stay an outer electron.

Also, with phosphorus, based on the given numbers, we apparently are not allowed to count the single proton to stay with the counting scheme of the Standard Model. Maybe we can only count the single proton if it is not involved in pulling in a quasi-inner electron. We will follow this path further.

And we notice something else: there is no element with 14 complete deuterons, we jump from 13 (silicon) to 15 (phosphorus). Structure-wise it would be a lithium nuclet on the right and a carbon on the left, resulting in 29 protons. It seems that the lithium nuclet initially does provide enough support to convert an outer electron to a quasi-inner one, but then cannot hold it in place, the structure on the lithium end of the nucleus is destroyed, decaying to silicon-29. So, this missing element with a deuteron count of 14 is most likely unstable, presumably with a very small half-life and—if created—another source for silicon-29. We have our first candidate for a missing element other than a noble gas.

Sulfur-32 and chlorine-35 (Fig. 4.10) are again very similar to silicon. For chlorine-35 we would expect 17 outer electrons according to the Standard Model and 17 deuterons and one PEP (= 18 inner electrons) following our definition. We see 12 + 11 + 5 + 5 +2 = 35 protons. We only see 6 + 5 + 2 + 3 = 16 deuterons, but there are two PEPs. The bluish carbon nuclet carries a single proton which provides the additional required outer electron. As (similar to silicon) the five-ending is back on the right, it is not possible to make that outer electron a quasi-inner electron. Instead, the PEPs provide the missing inner electrons for our count.

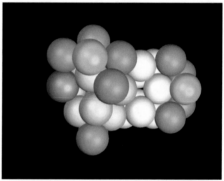

Figure 4.10 Chlorine-35 with an additional single proton (*brown*).

The same thing happens with argon-36, argon-38, and argon-40. For potassium-39 (Fig. 4.11) we would expect 19 outer electrons, 19 deuterons, and one PEP according to the Standard Model and our definition. We see 12 + 11 + 5 + 5 + 6 = 39 protons. We see 6 + 5 + 2 + 2 + 3 = 18 deuterons. This is the same case we saw before with silicon, with the five-ending to the right. The additional carbon nuclet is providing the single proton and therefore the missing outer electron.

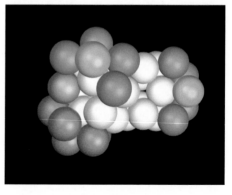

Figure 4.11 Potassium-39 with an additional single proton (*brown*).

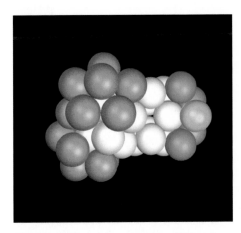

Figure 4.12 Calcium-40 with additional single proton marking (*brown*).

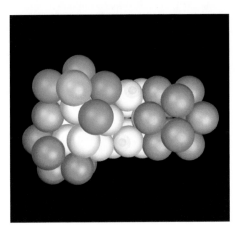

Figure 4.13 Scandium-45 with two quasi-inner electrons (*smaller yellowish spheres*) and two single proton markings (*brown*).

The next element is calcium-40 (Fig. 4.12). We would expect 20 outer electrons and 20 deuterons; no PEP according to the Standard Model and our definition. We see $12 + 11 + 6 + 6 + 5 = 40$ protons and $6 + 5 + 3 + 3 + 2 = 19$ deuterons. The carbon nuclet provides a proton and an additional outer electron. The five-ending to the right is still there, so no option exists to change the outer electron to a quasi-inner electron. Therefore, we arrive at 20 inner electrons and 20 outer electrons.

With scandium-45 (Fig. 4.13) we would expect 21 outer electrons, 21 deuterons/single protons, and three PEPs, equaling 24 inner electrons according to the Standard Model and our definition. We see $12 + 11 + 11 + 5 + 6 = 45$ protons. We also see $6 + 5 + 5 + 2 + 3 = 21$ deuterons, two protons, and one PEP which provide two additional outer electrons and one inner electron.

Now we have 23 outer electrons and 22 inner electrons with scandium. However, there is again the structure that pulls in outer electrons to the nucleus—this time two of them. This corrects the count to 21 outer and 24 inner electrons, two of them being quasi-inner electrons. Why two? With scandium, we see two additional carbon nuclets for the first time, so we have two single protons to cover. However, this is different from phosphorus, where the two endings on one carbon pulled in the outer electron. This time the outer electrons are pulled between whole branches, at least on one side. Also, we are missing an element again, this time with 20 deuterons. This missing element follows calcium. It is most likely the reason for the stability of calcium-46 and calcium-48, as a decay upward with a β– step, is not easy. We will look at this in more detail in Section 11.3.

For titanium-46 (Fig. 4.14) we expect 22 outer electrons, 22 deuterons, and two PEPs ($= 24$ inner electrons) according to the Standard Model and our definition. We see $12 + 11 + 11 + 6 + 6 = 46$ protons. We see $6 + 5 + 5 + 3 + 3 = 22$ deuterons but no additional PEPs. There are two single protons which provide two more outer electrons as part of the carbon nuclets. We end up for now with 24 outer electrons and 22 inner electrons.

Here we have again two structures on different branches which could pull in an outer electron each, changing them to quasi-inner electrons, which corrects the count, resulting in 22 outer electrons and 24 inner electrons (two of them quasi-inner electrons). We can think of this situation as the additional proton in the carbon structure connecting to a deuteron on another ending via the help of the quasi-inner electron, giving the previously outer electron two anchor points on the nucleus.

For vanadium-51 (Fig. 4.15) we expect 23 outer electrons and 23 deuterons and 5 PEPs, according to the Standard Model and our definition. We see $12 + 11 + 11 + 6 + 6 + 5 = 51$ protons as well as $6 + 5 + 5 + 3 + 3 + 2 = 24$ deuterons and one additional PEP. There is one single proton in the right bluish carbon nuclet. This provides one more outer electron. There is also an additional proton in the left carbon nuclet, providing an additional outer electron. For now, we end up with 26 outer electrons and 25 inner electrons.

With three structures to pull in outer electrons, converting them to quasi-inner ones, the numbers correct to 23 outer electrons and 28 inner electrons, 3 of them quasi-inner electrons. Two of them are the ones we have seen before, pulling between the branches on the top side. What is the third one? If there is enough branch structure available to the side, another option seems to be available to pull in outer electrons, in this case a third one, visible at the bottom. This would actually be a deuteron–deuteron connection. The structure of vanadium might be off because of this occurrence, this is kind of an anomaly and deserves more research.

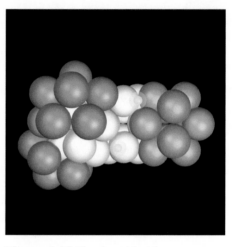

Figure 4.14 Titanium-46 with two quasi-inner electrons and two single proton markings.

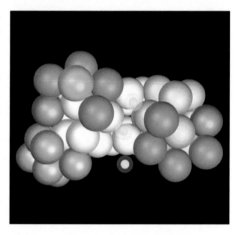

Figure 4.15 Vanadium-51 with three quasi-inner electrons and two single proton markings.

Apparently, we are also missing an element with 23 deuterons. For chromium-50 (Fig. 4.16) we would expect 24 outer electrons and 24 deuterons and 2 PEPs according to our rules. We see $12 + 11 + 11 + 8 + 8 = 50$ protons and $6 + 5 + 5 + 4 + 4 = 24$ deuterons. The one proton from the right bluish carbon nuclet provides one more outer electron. There is also an additional proton from the white, left carbon nuclet. The count for now is 26 outer electrons and 24 inner electrons.

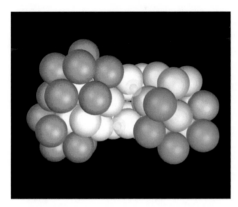

Figure 4.16 Chromium-50 with two quasi-inner electrons and two single proton markings.

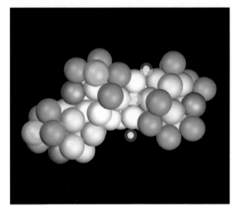

Figure 4.17 Copper-63 with three quasi-inner electrons and three single proton markings.

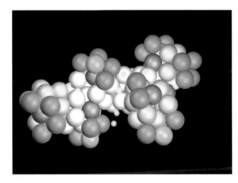

Figure 4.18 Silver-107 showing the quasi-inner electrons.

The structures again pull in two outer electrons between the branches, converting them to quasi-inner electrons changing the count to 24 outer electrons and 26 inner electrons, 2 of which are quasi-inner electrons.

The pattern is becoming clear; we move to copper-63 (Fig. 4.17) to see what happens when the next carbon nuclet appears. Copper is element 29, we therefore expect 29 outer electrons, 29 deuterons/single protons and 5 additional PEPs. We see $12 + 11 + 11 + 11 + 8 + 5 + 5 = 63$ protons and $6 + 5 + 5 + 5 + 4 + 2 + 2 = 29$ deuterons. Three carbon nuclets provide 3 additional single protons which brings the count to 32 outer electrons. With 2 additional PEPs we have 31 inner electrons.

However, again there are branches that would pull in outer electrons, changing them to quasi-inner ones. This brings the count to 29 outer and 34 inner electrons, three of them being quasi-inner electrons.

Silver-107 (Fig. 4.18) has 6 single protons but 8 quasi-inner electrons. The branches can carry some additional quasi-inner electrons on top of the single proton induced ones.

4.4.1 Quasi-inner electrons— what does it all mean?

The concept of an outer electron being pulled into the minimal bounding sphere (MBS) of the nucleus by the structure seems strange and alien at first. Those so-called "quasi-inner electrons" reside not inside the proton structure as the real inner electrons do, but just above it, that is, for nearly all cases between branches of the nucleus. As such they exert a pulling force between the protons and endings they connect to. Sometimes the force is too high and the nucleus disintegrates.

However, for all intents and purposes, especially mass calculations, we can count them as "inner electrons" too. In combination with protons they look like PEPs. This mechanism explains the "neutron"/proton ratio change with heavier nuclei which is observed. Those semi-PEPs do not negatively affect the stability of a nucleus as normal PEPs do. It is also clear that the number of outer electrons is no longer equal to the number of deuterons and single protons not bound in deuterons. We have to subtract the number of quasi-inner electrons that were sourced from the pool of outer electrons. This might also affect our understanding of the term "component" in the isomer definition. We might have to consider the quasi-inner electrons too.

> We learned in this section that our initial element number definition as deuteron plus single proton count (see Sections 14 and 4.3) was not 100% correct or rather incomplete as we found one more rule that completes this. We have to modify our definition of the element/atomic number as *the number of deuterons plus the number of single protons not able to pull in an outer electron to become a quasi-inner electron.* The number of outer electrons is defined as: *the number of deuterons plus the number of single protons minus the number of quasi-inner electrons.*

When we do so, we are initially more or less in lock-step with the numbering system of the Standard Model. Vanadium shows the first divergence between the two counting systems. However, there is still a slight question mark on the structure of vanadium. Manganese and gallium follow. Common to all those elements with divergences in this part of the PTE is that they do not show a wide range of isotopes. We also see some missing elements according to our definition. The first one we see together with silicon, the next two after calcium, one after scandium, and then again around those first diverging elements (Fig. 4.15). This leads to the hypothesis that those irregular elements (like scandium) are decay products of those unstable missing elements above and/or below the diverging elements. In case of scandium there is for example one unstable missing element (with element number 22) just above it with 45 protons too, which very likely is the "source" of scandium. The missing element is unstable, its quasi-inner electron rips off a proton from a lithium nuclet and creates a carbon nuclet from the boron ending. Now it can support two quasi-inner electrons, with one outer electron moving in a very similar manner to β+ decay. We have reached scandium-45.

With bromine the lock-step with the Standard Model is completely broken (see also Appendix I), never to be repaired again. According to our new counting definition, bromine is element 37 and krypton is element 38. Elements 35 and 36 are again missing. The difference trend gets larger as the nuclei get heavier. The repercussions for decay schemes will be considered in Sections 11.3 and 11.5.

The new SAM nuclear periodic table of elements in Appendix B (Figs. B.1/B.2) shows the application of our new element number rule all the way up to americium. We also see some interesting element reorders (see also Appendix I) as well as missing elements. These missing elements are of special interest, and we have already talked a little about their positioning. If we look at the structures, we often see a single nitrogen ending (two-ending) in those missing elements. This structure is apparently unstable, except for nitrogen itself. We see combinations of a carbon nuclet and a lithium nuclet, also unstable. However, some of the missing elements could be stable, we just have to look for them.

> There might still be new, undiscovered stable elements out there in addition to a multitude of unstable ones.

Another consequence will be to question the group categorization of some elements in the future. Based on the structure some currently missing elements might be a better fit than current assignments. However, that is a future research topic.

4.5 CHEMICAL BONDS

4.5.1 Ionic bonds

The simplest chemical bond we know is the ionic bond where valence numbers seem to apply. A good example, already mentioned, is table salt (NaCl or sodium chloride). According to textbooks sodium has a valence value of +1 and chlorine a value of -1. The + means that the sodium donates one outer electron to the chlorine which needs only one additional outer electron to appear like argon—in terms of its outer electrons.

In effect, we see that chemical bonds are—looking through SAM glasses—a connection formed by sharing electrons whereby both atoms are reduced to the nearest noble state, in other words, sodium to neon and chlorine to argon. We can therefore conclude that even though chemistry is the domain of the outer electrons, chemical bonds seem to be a continuation of the nuclear prevalence of a noble gas. A chemical bond mimics the noble gas state, which in turn is the lowest energy state. We see this also reflected in the size of atoms—the more we go to the right of the periodic table of elements (PTE), where the noble gases are located, the smaller the atom tends to be, relative to the cycle-of-eight. Therefore, the noble endings of the nucleus are easily recognizable as being correlated to the smallest atom sizes. This is an ongoing research topic.

Returning to NaCl bonds, we notice that the extra deuteron on the neon nucleus, making sodium, connects to the chlorine atom by connecting to the positive spots on the chlorine nucleus. This is logical, because otherwise the electron from sodium could not be shared and no chemical bond would exist. This is the case with noble elements (gases) as well. In SAM the completed capping phase is the noble configuration, or rather the last step in the cycle-of-eight. This configuration is a preferred state at the atomic level too. For chemical reactions, we can think of a densest packing rule at the atomic level—not at the level of the nucleus as is the case with nuclear physics. The theme for chemistry seems to be the same as for the nucleus: completing the geometry, getting closer together, although somewhat different and with electrons—not protons.

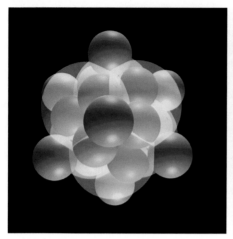 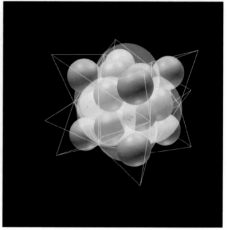

Figure 4.19 Carbon-20 shown with inner electrons twice the size of the protons.

Figure 4.20 Carbon-20 used to illustrate the directional component of the positive connection points of the carbon nucleus.

Depending on the element and maybe even on the isotope, the number of positive connection spots on the nucleus varies and influences the chemical bond, or oxidation state numbers. Since we show that in the 4th row of the PTE branches are growing in parallel on several active endings, we have a very dynamic mix of numbers.

Another research topic is that the endings are suspected of making connections to the parent nuclei instead of with different nuclei, resulting in more complex oxidation state values.

There are different types of chemical bonds, of course, but a discussion of them is outside the scope of this book. The metallic bond however will be discussed in Section 4.5.2.

One more topic we can talk about at this point: the *coordination number* of an atom is the number of atoms or ions bonded to it. It is also called *ligancy*. The ion or atom bonded to the central atom is called a *ligand*. The same is true at a higher level for molecules as well as crystals, although the calculation is different. The coordination number is determined by counting the other atoms to which the "central" atom is bonded. The common coordination numbers are 4, 6, and 8. Those common coordination numbers might just be related to the strong open connection points on the nucleus, as discussed in Section 4.1.

Looking at carbon and its isotopes and bringing this in the context of strongly versus weakly bound proton–electron pairs, we can see that the first eight proton–electron pairs are located on the most positive spots available where they can attach to the nucleus with their own carried electron (Fig. 4.19).

According to the model there are more locations (triangular spots) where a proton–electron pair could be forced to reside. However, the connection is not one made by sharing the inner electrons as is normally the case for the nucleus. Usually we see β– decay occurring under these circumstances. However, before that happens we have

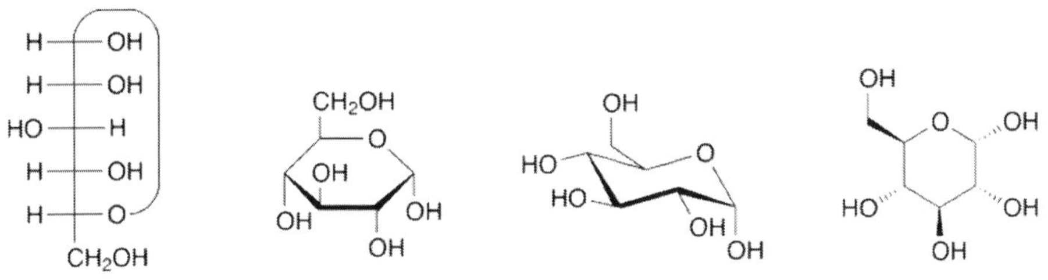

Figure 4.21 Molecules based on carbon and hydrogen.
[Wikipedia 2021/Glucose]

to recognize that there is a concept called the "neutron drip line." The nucleus is so much overloaded with extra proton–electron pairs that they simply drift out again. This known phenomenon can be explained through SAM. It is simply a logical threshold that we can point at or calculate as shown in Table 4.2.

In our example, carbon-12 is the base. For each and every proton–electron pair we add we see that the stability decreases. We also see that the carbon-20 shows the maximum of strongly bound proton–electron pairs, eight in total. Any proton–electron pair forcefully added after all spots have been taken (carbon-20) will not connect to the nucleus due to a lack of any positive spots. Such a PEP will be repelled easily.

Coming back to the coordination number of an atom—in this case carbon—the strongly binding positive spots also include a directional component as shown in Figure 4.20—assuming the spots are not taken by additional PEPs but can be used for ionic bonds.

Carbon is one of the best known elements, perhaps on par with oxygen, which we all need to breathe. Carbon is the fundamental building block for chemistry, at least in the field of organic chemistry. One of the most recognizable building blocks in biochemistry is the ring structure, such as in a glucose molecule. The "backbone" for these type of chemical compounds is made up of carbon atoms.

In more detail we see oxygen as an ending and all the left-over connections are covered by hydrogen atoms (Fig. 4.21). Oxygen in turn has only one side left for connections and can therefore be double-bound on one side, or in a tilted position, making one bond to the main structure and one to a hydrogen, the so-called OH bond. This is essentially the basis for all life, although greatly simplified. Looking at the carbon atom we can observe that it is like an oxygen without its capping deuterons. We see that carbon has a left and right side to its structure.

Carbon on its own is even more special. As we have seen, carbon is very interesting due to its perfect geometry. The chemical connection points are not yet in the distorted carbon shape, localized on one side of the nucleus, but the four chemical connection points are in a perfect tetrahedron shape and/or cubical shape, depending on the viewing angle. This in effect shows why a carbon can make chains. This can turn into the abovementioned carbon ring structure for glucose (Fig. 4.21).

4.5.2 Metallic bonds

Metallic bonding is a type of chemical bonding that results from the electrostatic attractive force between conduction electrons (in the form of an electron cloud of dislocated electrons) and positively charged metal ions. It may be described as the sharing of free electrons among a structure of positively charged ions (cations). Metallic bonding accounts for many physical properties of metals such as strength, ductility, thermal and electrical resistivity, conductivity, opacity, and luster [adapted from Wikipedia 2021/Metallic_bonding].

Examining the structure of the elements that are known for their strength, we can see that the greater the number of endings and the larger they are, the stronger the metallic bond. According to the description, the positive ions (metal) are bound together through dislocated electrons. That would also mean that the greater the number of connections a metal can make and the more directions these bonds can go (larger and more endings), the stronger a metal becomes.

Comparing sodium with magnesium, we now see why magnesium would be stronger and more rigid. It has two active lithium nuclets instead of one. Studying titanium, which is well known for its strength and lightness, we now recognize why this would be such a strong metal. It has two lithium nuclets and one carbon nuclet, in total three active endings. It is therefore able to make many connections in "all" directions resulting in a strong metal.

These correlations are made through the structure of the nucleus, now that we understand it. The reader can compare any element and its properties with any other element—with the help of the SAM nuclear PTE (Appendix B) and the element reference (Appendix H). Carbon is used to make iron stronger by creating an alloy of the two elements. A quick look tells us that carbon would be indeed very strong by being able to make four connections through eight positive spots with electrons—the diamond. So, in essence we put in the metal iron some attributes associated with diamond by replacing some iron atoms with carbon, making the iron therefore more rigid and strong, but also less pliable or flexible. These attributes and the technique of adding carbon to iron to make it harder have been well known since the dawn of the Iron Age.

The protection layers of most tools (e.g., hammers and screwdrivers) are made up of metals such as vanadium and chromium. Looking at those elements we see that they have a high oxidation state and are made up of several active endings and no noble parts. Multiple connections can be made in almost any direction; we see this reflected in the strength of the material. This is, of course, still a research topic.

4.6 SUMMARY

In this chapter we discussed further advancements of the model:

- We looked at the structure of the nucleus and possible strong or weak connection points for PEPs. There seems to be a correlation between the number and type of those connection points and the maximum number of "neutrons" an element can handle (neutron drip line).

- We can now discern three types of isomers (component, ending type, base structure).
- We explained the "neutron"/proton ratio through the introduction of quasi-inner electrons, located between branches and just above the endings on active carbon nuclets.
- We introduced a new system to number the elements based on the deuteron plus the single proton count, only counting single protons unable to pull in an outer electron to become a quasi-inner electron.
- When using this new counting system a lot of spots open up which are not yet occupied by known elements.
- The shape of the nucleus and the positioning of the protons and inner electrons dictate the positions of the outer electrons.
- The positioning of the outer electrons determines the angles of possible connections to other atoms. This demonstrates the connection to chemistry through ionic and metallic bonds.

CHAPTER 5

Other known aspects of elements

5.1 ELECTRICAL CONDUCTIVITY

Here we have a list of metals in order of decreasing electrical conductivity. Consulting the Atom-Viewer we can visualize the elements (Figs. 5.1–5.8).

Interestingly, it is obvious that the first three best conducting elements are also the known noble metals, that is, silver (Ag), copper (Cu), and gold (Au).

The other elements are not so clear in sharing similar structures. We can conclude that the more noble the metal seems to be, the better the element is able to conduct electricity.

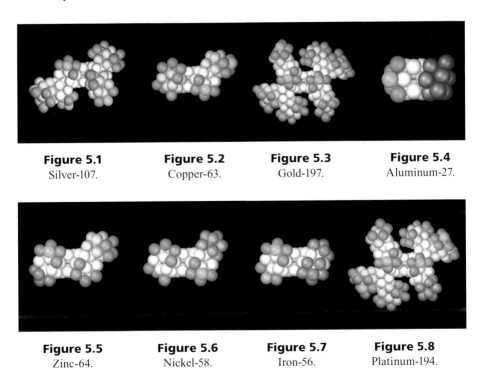

| **Figure 5.1** | **Figure 5.2** | **Figure 5.3** | **Figure 5.4** |
| Silver-107. | Copper-63. | Gold-197. | Aluminum-27. |

| **Figure 5.5** | **Figure 5.6** | **Figure 5.7** | **Figure 5.8** |
| Zinc-64. | Nickel-58. | Iron-56. | Platinum-194. |

5.2 ELECTRON AFFINITY

The electron affinity of an atom or molecule is defined as the amount of energy *released* when an electron is attached to a neutral atom or molecule in the gaseous state to form a negative ion. Equivalently, electron affinity can also be defined as the amount of energy *required* to detach an electron from the atom while it holds a single, excess electron thus making the atom a negative ion.

To use electron affinities properly, it is essential to keep track of sign. For any reaction that *releases* energy, the *change* ΔE in total energy has a negative value and the reaction is called an exothermic process. Outer electron capture for almost all non-noble gas atoms involves the release of energy and thus are exothermic. It is the word "released" within the definition "energy released" that supplies the negative sign to ΔE.

Although electron affinity varies greatly across the periodic table of elements, some patterns emerge. Generally, non-metals have more positive electron affinity than metals. Atoms whose anions are more stable than neutral atoms have a greater electron affinity. Chlorine most strongly attracts extra electrons; neon most weakly attracts an extra electron. The electron affinities of the noble gases have not been conclusively measured, so they may or may not have slightly negative values (Fig. 5.9). Some values in the list are apparently estimates (Table 5.1).

We do not yet know how electron affinity relates to the structure of the nucleus, the inner electrons, and the outer electrons. Nitrogen-14 is a very interesting case—an ongoing research topic. The relative numbers for noble gases and elements close to this state with valence equal to -1 are to be expected. Electron affinity is also helpful when explaining behavior in experiments.

Let us consider an example: assume we have an experiment with a cathode and an anode in a chamber. The chamber can be evacuated or filled with various gases. Gases can be introduced through the anode too. In this setup, nuclear reactions on the anode and/or the cathode, as well as chemical reactions in the atmosphere, can be tested. In the case where we fill the chamber with oxygen and push hydrogen through the anode in an electrical environment, the oxygen will be negatively ionized according to its electron affinity. It will travel toward the anode where it will meet the positively

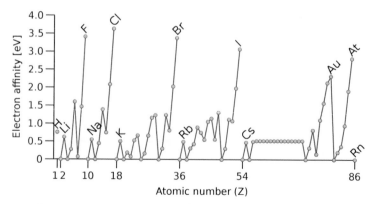

Figure 5.9 Electron affinity (eV) shown against atomic number.
[Karjono 2012]

Table 5.1 Electron affinity (eV) of some isotopes.

Isotope	Electron affinity (eV)	Remarks
Hydrogen-1	0.754195	
Hydrogen-2	0.75459	
Helium-4	−0.5	Estimate
Lithium-7	0.618049	
Beryllium-9	−0.5	Estimate
Boron-10	0.279723	
Carbon-12	1.2621226	
Carbon-13	1.2621136	
Nitrogen-14	−0.07	
Oxygen-16	1.4611136	
Oxygen-17	1.461108	
Oxygen-18	1.461105	
Fluorine-19	3.4011898	
Neon-20	−1.2	Estimate

ionized hydrogen (protons). The result will be that hydrogen nuclei pushed through the anode are used to create water, preventing any possible nuclear reaction between the anode material and the hydrogen nuclei (protons). Fluorine and chlorine should have even bigger dampening effects than oxygen according to their electron affinity. Other gases with less electron affinity will not prevent this possible nuclear reaction. Nitrogen is one example which will not prevent a reaction as are all the noble gases. They could help stimulate the formation of an electrical environment that takes part in or at least supports the nuclear reaction. Interestingly those gases are what are commonly known as shielding gases (e.g., in welding). Such gases prevent chemical reactions, but in an electrical environment they might welcome nuclear reactions.

5.3 NUCLEAR MAGNETIC MOMENT AND SPIN

The magnetic moment is the magnetic strength and orientation of an object that produces a magnetic field. This includes permanent magnets, moving fundamental particles, atoms, and various molecules. The term magnetic moment normally refers to a magnetic dipole moment. Higher order terms (e.g., magnetic quadrupole moment) may be needed in addition to the dipole moment for extended objects (in size). The magnetic moment of the outer electrons far exceeds the magnetic moment of the nucleus. Once full atoms are used to determine the magnetic moment, the moment of the nucleus can be neglected.

Spin is defined as an intrinsic form of angular momentum carried by fundamental particles, groupings of fundamental particles, and whole atomic nuclei. When certain fundamental particles move through a magnetic field, they are deflected in a manner that suggests they have magnetic properties (see also Section 5.6). Spin itself is a bizarre physical quantity: it is analogous to the spin of a planet in that it provides angular momentum and a tiny magnetic field which produces a magnetic moment. However, based on current models it is agreed that fundamental particles do not rotate in this way. Therefore, spin must be an *intrinsic property* of a fundamental particle.

A magnetic moment is a vector quantity, and the direction of the magnetic moment is defined by the spin of fundamental particles—according to the Standard Model. The problem is that only one direction of the vector that describes the direction of the magnetic moment can be measured at the same time—following Heisenberg's Uncertainty Principle. And the direction can only have discrete values. However, fortunately, according to the Standard Model, spin is the same in all three dimensions. So, there is just one spin number defining the directional vector of the magnetic moment that can be assigned to fundamental particles, groupings of fundamental particles, and whole atomic nuclei. As a result, the directional vector of a magnetic moment in the Standard Model is limited to a few distinct values.

An electron's magnetic moment is $-9.284764620 \times 10^{-24}$ J/T (Joule/Tesla). Electron spin has only one possible value: ½. Spin is the orientation of the magnetic moment, a vector quantity. The quantum spin number is also written as +/- ½ or up ½ and down ½. The torque on the electron resulting from an external magnetic field depends on its orientation with respect to the field (right-hand rule). A proton is ascribed a spin of ½ and a magnetic moment of $1.4106067873 \times 10^{-26}$ J/T. The torque on the proton resulting from an external magnetic field is toward aligning the proton's spin vector in the same direction as the magnetic field vector. The "neutron" is assigned a spin of ½ and a magnetic moment of $-9.6623647 \times 10^{-27}$ J/T. The torque on the neutron resulting from an external magnetic field is toward aligning the neutron's spin vector opposite to the magnetic field vector.

According to the Standard Model, protons tend to form pairs of opposite total angular momentum. The same goes for "neutrons." Therefore, the magnetic moment of a nucleus with even numbers of both protons and neutrons is zero, while that of a nucleus with an odd number of protons and even number of neutrons (or vice versa) will have to be that of the remaining unpaired nucleon (Table 5.2). The pairing idea was developed while scientists were looking at the first elements. Another "observation" is that spin is integral for nuclei of even mass number and half-integral for nuclei of odd mass number.

What do we see when we look at the structure of an atom in SAM and its spin? A deuteron would be like a bar, or a stick, with two charged poles. So, in an external field we would expect a rotational component (it turns), also known as magnetic nuclear resonance. A helium nucleus would have two of those perpendicular to one another, canceling out the moment due to the perfect shape of the tetrahedron. The same thing is true for the carbon nucleus in the form of an icosahedron. In other words, two paired

Table 5.2 Various magnetic moments and spin. [IAEA 2021]

	Magnetic dipole moment	Spin
Electron	$-9284.764 \times 10^{-27}$ J/T	1/2
Proton	2.79284734462 μ_N $14.106067 \times 10^{-27}$ J/T	1/2
Neutron	-1.91304272 μ_N -9.66236×10^{-27} J/T	1/2
Deuteron (hydrogen-2 nucleus)	0.857438228 μ_N $4.3307346 \times 10^{-27}$ J/T	1
Triton (hydrogen-3 nucleus)	2.97896244 μ_N $15.046094 \times 10^{-27}$ J/T	1/2
Helion (helium-3 nucleus)	-2.12749772 μ_N $-10.746174 \times 10^{-27}$ J/T	1/2
α particle (helium-4 nucleus)	0×10^{-27} J/T	0
Lithium-6	0.82205667 μ_N	1
Lithium-7	3.2564268 μ_N	3/2
Beryllium-9	-1.1778 μ_N	3/2
Boron-10	1.80064478 μ_N	3
Boron-11	2.6886489 μ_N	3/2
Carbon-12	0 μ_N	0
Carbon-13	0.7024118 μ_N	1/2
Carbon-14	0 μ_N	0
Nitrogen-14	0.403761 μ_N	1
Nitrogen-15	-0.28318884 μ_N	1/2
Oxygen-16	0 μ_N	0
Oxygen-17	-1.89379 μ_N	5/2
Oxygen-18	0 μ_N	0
Fluorine-19	2.628868 μ_N	1/2
Neon-20	0 μ_N	0

deuterons seem to have negated any moment. Carbon is three of those paired deuterons in three dimensions.

Often the magnetic dipole moment is given in nuclear magnetons (μ_N); 1 nuclear magneton equals $5.050783699 \times 10^{-27}$ J/T.

Spin is based on an independent particle model and quantum mechanics, simply adding up the values is possible there. SAM is no such model, we therefore consider the spin of the Standard Model to be irrelevant for SAM.

There is however a magnetic moment to a nucleus and a whole atom with direction and value and there is torque when it is exposed to a magnetic field—nuclear magnetic resonance imaging (NMRI) is a technique making use of that process. A nucleus is excited by NMRI which applies torque via a magnetic field and creates a magnetic moment in atoms with chemical connections. When released, the nucleus resettles to its original position, in doing so highlighting the structure of the outer electrons connected to the nucleus. This resettling yields energy that is used in the MRI machine to create an image.

In SAM, we would expect there to be a magnetic moment on fundamental particles and atoms, which has a value and a direction, based both on the structure of the nucleus and the distribution of outer electrons, which itself depends on the structure of the nucleus. The direction of the magnetic moment would not be limited to a few very discrete values that can be described with one number. It should not be impossible, based on the available structural information of an atomic nucleus and the positioning of the outer electrons, to fully determine the direction and the value of the magnetic moment. Ending type isomer and component isomer configurations of an isotope would however show a slightly different magnetic moment and possibly a slightly different direction. If we therefore expose different isomeric configurations or isotopes of an element (e.g., silver) to a magnetic field they might behave slightly different. We also have to consider the option that an experimental setup using electromagnetic forces changes a nucleus, at least temporarily. What we can already see is that the magnetic moment, as well as the spin of elements that are structurally symmetric, is usually close to zero. Details on the magnetic moment of a nucleus are again an ongoing research topic for the SAM team. The same is true for parity. Parity is another intrinsic property of fundamental particles—according to the Standard Model. Based on those, the parity of atoms and molecules can be calculated. As with spin we will have to look in the future at the experiments that forced the Standard Model to introduce these properties into their theories and see what we can conclude from those experimental results with SAM in mind.

5.4 THE SHELL MODEL OF CURRENTLY ACCEPTED THEORY

In stark contrast to SAM the Standard Model organizes outer electrons in shells and orbitals (Fig. 5.10). Each shell is composed of one or more sub-shells, which are themselves composed of atomic orbitals. For example, the first (K) shell has one sub-shell, called 1s which can hold two electrons; the second (L) shell has two sub-shells, called

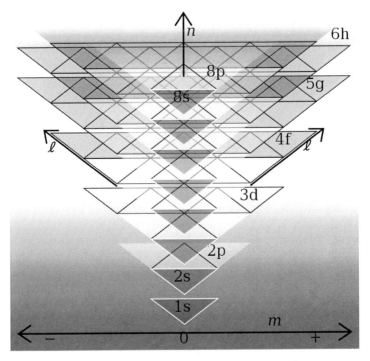

Figure 5.10 Representation of the shell model.

[Wikipedia 2021/Electron_shell]

2s and 2p (which can hold six electrons); the third (M) shell has 3s, 3p, and 3d (which can hold ten electrons); the fourth (N) shell has 4s, 4p, 4d and 4f (which can hold fourteen electrons); the fifth (O) shell has 5s, 5p, 5d, and 5f and can theoretically hold more in the 5g sub-shell (which can hold eighteen electrons) that is not occupied in the ground-state electron configuration of any known element. No known element has more than 32 electrons in any one shell.

In this model in the ground state of an atom or ion, electrons fill atomic orbitals of the lowest available energy levels before occupying higher levels (Fig. 5.10). For example, the 1s sub-shell is filled before the 2s sub-shell is occupied. This way, the electrons of an atom or ion form the most stable electron configuration possible. Another rule of the model states that if multiple orbitals of the same energy are available, electrons will occupy different orbitals singly before any are occupied doubly.

Hydrogen starts the 1s orbital and helium closes it; lithium starts 2s and beryllium closes it; boron opens 2d and neon closes it; sodium opens 3s and magnesium closes it; aluminum opens 3p and argon closes it; and potassium opens 4s and calcium closes it. One would now expect scandium to open 4p, but instead 3d is opened. Zinc closes 3d, gallium finally opens 4p, and krypton closes 4p. The argument is that the 3d orbital is less energetic than 4p. The same thing repeats with the rare-earths which open and close 4f after 6s, then open and close 5d until it continues with 6p. And again more theoretically with 7s, 5f, 6d, and 7p.

There are 10 elements among the transition metals and 10 elements among the lanthanides (rare-earths) and actinides for which the predicted electron configuration differs from the one determined experimentally using the shell model interpretation. The shell model reveals itself to be a game of numbers that fits more or less what is found in nature. Wouldn't it be better to understand first what is going on? For example, in copper the 4s orbital should be occupied fully before the 3d orbital according to the rules. However, the "measured" electron configuration of the copper atom in the interpretation of the shell model shows the 3d orbital to be fully filled first.

From the perspective of SAM we do not have shells or orbitals in the realm of the outer electrons. Instead, we have different active endings on the nucleus that have not been made chemically inert which interact with some of the outer electrons and represent connection points for bonds of various kinds. The orbitals are in essence a remaining piece of the cycle-of-eight idea taken beyond the point of its breakdown—the transition metals. In SAM, the structure is what it is and what the active endings determine it to be, it cannot be forced into shells and orbitals guided by magic numbers.

The reason the orbital system goes wrong is becoming clear now. The 2s and 2p orbital in the 2nd row represents the completion/capping of a carbon nuclet (becoming the noble gas neon). The completion of the 3rd row is still one carbon nuclet (argon). The completion of the 3rd and the 4th carbon nuclet does not happen in sequence due to the parallel growth of three branches (three carbon nuclets in the making). That is why it was decided in the Standard Model that the 4p orbital for iron is not filled first, rather the 3d orbital is. Otherwise, a noble gas would appear there and not iron. This means we can show with the help of structure (SAM accepted as truth) why the current orbital system breaks down.

Copper had issues in the shell model (3d unexpectedly filled before 4s) too. Let us now consider copper in SAM. Copper-63 in SAM has two inert backbone nuclets and two carbon nuclets. The right one has two four-endings and can also be considered inert. There is one five-ending on one backbone nuclet and one five-ending on the left active carbon nuclet. The frames visible in Figure 5.11 show the open connection points, not covered by a four-ending. That left side is very oxygen-like. Overall, copper looks to be close to an unknown noble element with four completely capped carbon nuclets. We would therefore expect 3d to be filled before 4s.

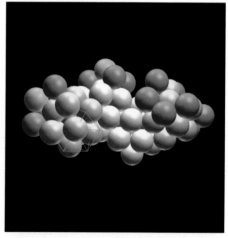

Figure 5.11 Copper-63 with internal connection frames displayed.

And there is one other conclusion after looking at the shell model, that is, that the quantum numbers of the currently accepted theory are a substitute for the structure of the nucleus. Once we have it—and we think we do—we can do away with those magic numbers and replace them with something less mystifying.

5.5 ARE THERE "SHELLS" IN SAM?

Neon becoming noble as an example means that the full ending/noble ending cannot react any further. A complete electron shield is now around the nucleus, no more connection points are available. The state of a full ending on elements such as neon is equivalent to the so-called complete orbital and the completed shell of the valence–shell model. One could equate the completed shell with the state of the completed ending, but we refrain from that for now. First, we need to know more about the outer electrons and their positions. Also, there are no shells within shells as the nucleus grows differently in SAM. The nucleus is based on the backbone of inert carbon nuclets, creating an elongated, branched form.

Additionally, there is talk about so-called "core electrons" in the Standard Model. Those are outer electrons that are not valence electrons. Only the valence electrons participate in chemical bonds, which are the outer electrons of the active endings. The core electrons, from SAM's viewpoint, are the outer electrons related to capped carbon nuclets. They are neither inner electrons nor quasi-inner electrons.

5.6 PARAMAGNETISM, DIAMAGNETISM, AND FERROMAGNETISM

Magnetism is an interesting effect since it is created at the atomic level. Best known are the permanent magnetic materials: iron, cobalt, and nickel. However, we will first look at paramagnetism:

Paramagnetism is a form of magnetism whereby certain materials are weakly attracted by an externally applied magnetic field, and form internal, induced magnetic fields in the direction of the applied magnetic field. [Wikipedia 2021/ Paramagnetism]

Paramagnetic materials are not permanently magnetic since they lose their magnetic effects once the external magnetic field is gone. Examples are O_2 and such metals as tungsten, aluminum, platinum, and chromium. Paramagnetic materials are attracted by a magnetic field. Diamagnetic materials are similar to paramagnetic materials but repelled by a magnetic field:

Diamagnetic materials are repelled by a magnetic field; an applied magnetic field creates an induced magnetic field in them in the opposite direction, causing a repulsive force. In contrast, paramagnetic and ferromagnetic materials are attracted by a magnetic field. [Wikipedia 2021/Diamagnetism]

Examples of diamagnetic materials are copper, silver, gold, lead, silicon, hydrogen, and nitrogen. Ferromagnetic materials are attracted to a magnetic field like paramagnetic materials, but they retain their magnetic properties after a magnetic field is gone:

Ferromagnetism is the basic mechanism by which certain materials (such as iron) form permanent magnets, or are attracted to magnets. [Wikipedia 2021/ Ferromagnetism]

The Standard Model attributes the magnetic effects to motion of the outer electrons. The circular motion around the nucleus creates an orbital magnetic moment. Additionally, the rotation of the electrons itself creates a spin magnetic moment.

As described in Sections 2.15 and 5.3–5.5 in SAM the structure of the nucleus dictates the positioning of the outer electrons. For heavier nuclei there is most likely no rotation of the outer electrons relative to the nucleus. They are kept in place over the connection points of the nucleus. However, the distribution of the outer electrons, depending on the structure of the nucleus in itself, creates in some cases a magnetic moment such as that exerted on a bar magnet.

If we look at oxygen, we see an uneven distribution of the inner electrons, which will create an uneven distribution of the outer electrons. This in turn will create a magnetic moment based on the distribution of the outer electrons—even for molecular oxygen (O_2)—when exposed to a magnetic field. For diamagnetic materials we have to assume that the resulting magnetic moment is not very big and points in the other direction. With ferromagnetic materials, the atoms are locked unidirectionally into a lattice grid that does not vanish once a magnetic field is removed.

5.7 SUMMARY

In this chapter we have considered several other aspects of the elements:

- Electrical conductivity.
- Electron affinity.
- Nuclear magnetic moment and spin.
- Magnetism.

All those aspects will hopefully at some point be related to the structure in some way, although currently the exact path is not yet clear to us. However, in each case at least a direction to move forward is visible. Also, we looked at the shell model and how it would be represented in SAM.

5.7.1 SAM and the pillars of observation

The main observational factors used in determining the structure of the nucleus with SAM and therefore developing SAM were:

- the "neutron"/proton ratio;
- nuclear reactions;
- isotopes and stability;
- the valence/oxidation state; and
- abundance data.

SAM is therefore in line with experimental and observational data about the atoms, as published for example by the IAEA.

The main decision reached while developing SAM was to remove the "neutron" as a fundamental particle and to replace it with the PEP (the proton–electron pair), thereby reintroducing electrostatic attraction into the nucleus. "Inventing the neutron" was *the catastrophic failure* of the Standard Model in our opinion. A lot of what went wrong and what remains still unexplained can be traced back to that decision.

As stated before, when we discussed "roads not taken" (Section 2.18), whenever there were options we chose the one that kept us closer to observation—not necessarily current models. SAM is about a clean slate, and we can therefore not avoid asking controversial questions.

CHAPTER 6

Interim conclusions

Where do we stand at the end of Part I (i.e., Chapters 1–5) of this book? So far we have:

- redefined the "neutron" as a proton–electron pair;
- provided a fixed structure of the nucleus of the atom—it represents a 3D fractal growth structure;
- provided rules for the buildup of the elements;
- defined the quasi-inner electrons and their location;
- created a new numbering system for the elements;
- put the inner electrons in relation to the outer electrons; and
- brought nuclear physics and chemistry closer together again.

As we already mentioned in the introduction, this is a work in progress. There are a lot of open research topics related to SAM, which we will summarize at the end of the book—Appendix E (Research topics). However, this is where observation and logic have led us so far. We are aware that this path is radically different from the path nuclear physics has taken in the last 100 years. However, if you feel that nuclear physics and its relation to chemistry is not a done deal and that it should be possible to turn away from a path that leads nowhere in our opinion, then follow us on our journey in Part II.

6.1 CHEMISTRY AND NUCLEAR PHYSICS

Chemistry overall appears to be a continuation of nuclear principles but at a lower level due to the distance of the outer electrons from the nucleus—about 10,000 times the diameter of the nucleus.

Chemistry is based on the sharing of electrons, but the receiving atom (capping phase) must have positive spots available, it cannot be noble. The number of positive spots seems to be the true limitation of the valence or oxidation values. Or stated differently: chemistry puts the elements involved into a "noble state" by sharing electrons.

The structure of the nucleus and its outer electrons as defined by SAM explains valence/oxidation state, chemical bonding, melting points, and other chemical properties of atoms as well as molecules.

Chemical bonds appear to do the "same thing" as the nucleus: densest packing with whole atoms creating molecules—instead of protons and inner electrons creating a nucleus. Just on another level. However, the number of connection points is different here than in the nucleus. The system is more complex in chemistry due to the more than 90 building blocks instead of only protons and electrons.

The preliminary conclusion we can draw here is that the chemical connection points are in a localized position of the structure of the nucleus and this is of interest for chemistry. The shape of the outer electron positions, or rather the angles of chemical bonds, are in part due to the structure of the nucleus. When looking at the structure in SAM, the higher the oxidation number and/or number of active endings, the more they are spatially distanced resulting in multiple different directions of connection potential and the stronger the element tends to be bound.

6.2 THREE CLASSES OF ELECTRONS

Outer electrons are acknowledged by the Standard Model as well as SAM. By declaring the "neutron" to actually be a proton–electron pair (PEP), a second class of electrons was introduced in SAM. Those are the inner electrons, which keep the protons of the nucleus together, acting much like glue.

A third class, the quasi-inner electrons, was introduced while looking at the issue of the "neutron"/proton ratio and trying to figure out a new—more appropriate—numbering system for elements. The quasi-inner electrons are located very close to the nucleus—between strong enough endings of a branch that can hold them in place (e.g., a five-ending on one side is not enough) and between the branches themselves.

6.3 NEW ELEMENT NUMBERING SYSTEM

Our revised numbering system is no longer based on the number of outer electrons. It is instead the number of deuterons plus the number of single protons unable to pull in an outer electron to become a quasi-inner electron. The structure reveals a lot of different sub-structures in various combinations, all with the same number of outer electrons, but with different properties. Changing to the new scheme is a step forward as it opens up some missing element spots, but still we cannot capture component isomers with this new system.

6.4 UNKNOWN ELEMENTS MISSING IN THE CURRENT PTE

Once you apply the new numbering system, a whole new world of new elements opens up. One class of elements are the unknown elements by missing element number. The other class would be component isomers or ending type isomers to existing, known, and also unknown elements. Most of these elements will be unstable with very short half-lives. The element technetium (unstable, artificially created) is probably a glimpse into this class of unstable elements.

6.5 WHAT IS STILL TO COME?

First, we have some open questions to address, not answered in Part I. We must:

- Review other models of the atomic nucleus.
- Address objections to a nucleus made of protons and electrons instead of protons and neutrons.
- Consider what the nature of spontaneous radioactive decay is. Is it really intrinsically random or is it an artifact of a model of the nucleus without structure?
- Revisit nuclear reactions in light of what we have learned so far.
- Consider the binding energy/mass defect of the nucleus. Is it possible to determine the binding energy of a nucleus with SAM based on the structure?
- Discuss the more complex nuclear reactions like fission and fusion. Can SAM provide insight into the cause of the asymmetric breakup of fissile isotopes for example?
- Consider how much of conventional wisdom about the elements we really have to follow. We already created a new numbering system and identified missing elements. We have to keep an open mind and assume no knowledge to be untouchable.

Then, of course, there are some new topics, still to be addressed. For example, we carefully avoided the term "transmutation" in Part I, but there really is no way to avoid that topic. In order to talk about it properly, we needed the buildup of Part I, but also nearly all of Part II will be required reading too. And then there is the topic of LENR (Low Energy Nuclear Reactions), also known as cold fusion or Condensed Matter Nuclear Science (CMNS). This is a real minefield, not only because "currently accepted theory" does not accept those reactions, because they are not possible in their theoretical framework, but also because of a wide range of contradicting opinions about what is being observed within the field of LENR itself. From our point of view, the issue of a missing theory (related to a missing model) is the biggest contributing factor to these disagreements. And a theory requires a working model.

We are aware this is controversial material since it will possibly break long-held belief systems. However, paradigms must be breakable, even if it is a difficult process, when all the evidence points in one direction. Observation always wins in the end.

PART II

"Systems, scientific and philosophic, come and go. Each method of limited under-standing is at length exhausted. In its prime each system is a triumphant success: in its decay it is an obstructive nuisance."

Alfred North Whitehead (1933), Adventures of Ideas

CHAPTER 7

Views of the Standard Model

7.1 OTHER MODELS OF THE ATOM

In his book, *Models of the Atomic Nucleus* [Cook 2010], Norman Cook provides a chronological classification of the various nuclear models developed since the 1930s. Figure 7.1 is an adaptation of Cook's fig. 3.3.

To show SAM's relation to these existing nuclear models we have placed it at the bottom of Figure 7.1 on the right-hand side in its own category "unforced models," because although the nucleus has structure, the nucleons are not forced into a lattice or something similar. As mentioned before, we see the electrons acting as "glue" between the protons in combination with the principle of spherically dense packing, manifesting overall as a centripetal force in each carbon nuclet with endings as well as for the whole nucleus.

The major types of nuclear models and their underlying assumptions are summarized in Table 7.1. The assumptions range from the nucleus being a gas, to a liquid, to a solid, and in a crystal lattice. Most of the models assume that the nucleons are moving around with varying degrees of mobility and are subject to the so-called Copenhagen interpretation of the Heisenberg Uncertainty Principle. It is Cook's contention that this assumption is unwarranted, given the known size of the nucleons and the nucleus, which are too close together to leave room for free motion in the sense of allowing a plausible mean free path, before running into each other (Section 7.2).

To illustrate the dogma imposed by the Uncertainty Principle, here is a quote by Dr. Cook regarding the notion of a fixed structure for the nucleus such as the FCC (Face Centered Cubic) lattice created by Cook himself or SAM:

> . . . *important developments over the last decade or so by Ulf Meissner in Germany, who's worked on something called nuclear lattice effective field theory (NLEFT). What's interesting about his work is that he is drawing conclusions about nuclear structure, at a level of spatial detail that is forbidden by the Uncertainty Principle and this is something that he does quite successfully and quite frequently in publications.*
>
> *It's interesting because it does contradict the dominant Copenhagen tradition in Quantum Mechanics which was originated by Bohr and supported by Heisenberg*

and Pauli and fought against by Einstein and others and is still an unresolved issue. But nonetheless, it has been a dominant view which is called the Copenhagen interpretation. You're probably all familiar with the Uncertainty Principle. If you localize a particle in space, that localization itself cannot be measured simultaneously with equal precision of, for example, the particle's momentum. There is a trade-off there.

I think the Uncertainty Principle itself is widely accepted by Einstein and others, but the interpretation is quite different considering what the particles' actual properties are, relative to what our understanding of our knowledge of the particles is. So, going back a few decades looking at lattice possibilities for nuclear structure, I wrote up a paper in the '70s, sent it off to a journal and the referees' comments were, well 'this is inconsistent with Quantum Mechanics, it's a non-starter. A liquid nucleus, a gaseous nucleus, a clustered nucleus is fine, but a lattice demands localization of the particles and therefore, cannot exist'. . .

[Cook & Di Sia 2018]

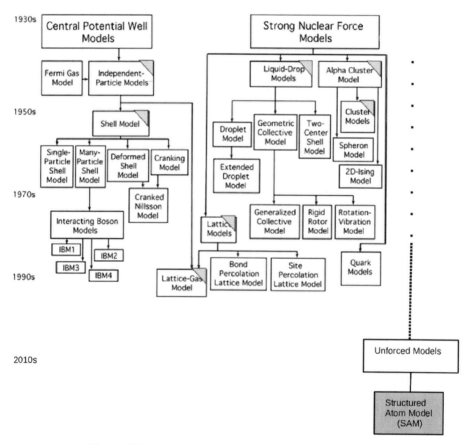

Figure 7.1 A chronology of the evolution of nuclear models.
[Adapted from Cook 2010, fig. 3.3]

Table 7.1 Models of the atom. [Adapted from Cook 2010, table 3.1]

Model	Type	Assumptions
Independent particle nuclear models (IPM)		Nucleons move nearly *independently* in a common potential.
Strong interaction models (SIM)		Nucleons are *strongly coupled* to each other because of their strong and short-range interactions.
Liquid drop model	SIM	The nucleus is regarded as a *liquid drop* with nucleons playing the role of molecules.
Shell model	IPM	Nucleons move nearly *independently* in a common static spherical potential which follows the nuclear density distribution.
Fermi gas model	IPM	Nucleons move *approximately independently* in the nucleus and their individual wave functions are taken to be plane waves.
Potential-well model		The nucleus is regarded as a simple real potential-well.
Alpha particle model	SIM	Alpha particles can be regarded as *stable sub-units* inside the nucleus.
Cluster model	SIM	The bulk of the nuclear binding energy is due to the binding of nucleons into *alpha particles*—with much weaker binding of the alphas to each other or to the small number of nucleons not contained within alphas.

According to Cook's review of the field, at least 37 nuclear models were employed by theorists and experimentalists as of 2010, where:

> . . . *each provides some insight into nuclear structure or dynamics, but none can claim to be more than a partial truth, often in direct conflict with the partial truths offered by other models.* [Cook 2010, 5]

Cook thinks it strange that since the 1930s liquid, gas, and cluster models were widely (and unsuccessfully) scrutinized and explored, but solid state lattice models were not very seriously considered during all these decades. Cook considers the Face Centered Cubic (FCC) model that he advocates to be more than "yet another nuclear model with only limited applications . . . rather the lattice is a strong candidate for unifying the many previous and seemingly-contradictory models employed since the 1950s"— arguing that the FCC replicates many of the properties of a nucleus because the relative positions of the nucleons (protons and neutrons) are fixed and known and are thus amenable to calculating the nuclear properties:

> *In the fission of palladium using the FCC model, we start with a default structure for each of the palladium isotopes. The core region (^{40}Ca) is assumed to be fixed*

with no movement of nucleons from their default positions, whereas the skin region consists of a sufficient number of populated FCC lattice sites that 46 protons and 56–64 neutrons are present (Fig. 11.30) [Fig. 7.2 in this text]. If the nuclear force were known for certain, it would be possible to ascertain which of the many alternative structures for any given isotope are energetically favored; lacking such certainty, a simulation can be undertaken where alternative structures are built and their relative stability is estimated from plausible assumptions about the nuclear force. [Cook 2010, 264–265]

So, here is palladium depicted as having at its core the structure of calcium-40, surrounded by a fluid skin with protons and neutrons in undetermined (to be discovered) lattice sites (Fig. 7.2). This is in contrast with the SAM depiction of palladium where, as for all larger atoms, we have a "neutralized/inert backbone" made from carbon nuclets (Fig. 7.3). Thus, we might say that the FCC model is somewhat of a hybrid itself, with a fixed core and a "fuzzy" periphery, whereas SAM is a solid structure all the way.

FCC still has vestiges of magic number theory, which breaks down as the atomic number increases:

Judging from the stable isotones, we find magic numbers at 20, 28, and 82 neutrons, but the peak at 50 is no greater than the peaks at 58, 78, and 80, and there is no indication of magic stability at 2, 8 and 126. Judging from the stable isotopes, we get magic numbers at 20, 28 and 50, but other small peaks arise at various numbers not normally considered to be magic: 54, 70, 76, and 80. [Cook 2010, 30]

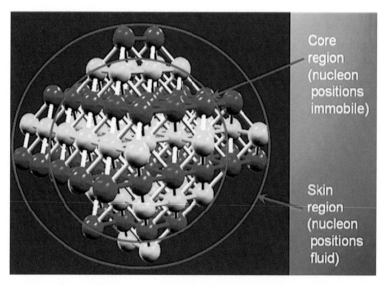

Figure 7.2 Palladium modeled in FCC.
[Cook 2010, 265]

In SAM, we know why this breakdown occurs because the structure starts developing branches as the various carbon nuclets are completed. We conclude this section by summarizing similarities and differences between the SAM and FCC models:

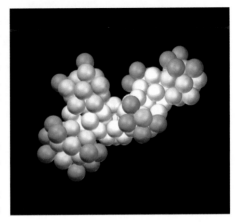

Figure 7.3 Palladium-104 nucleus modeled in SAM.

1 Both postulate that discovering the *structure* of nuclei is key to understanding their properties.

2 Both stipulate that for stability, no strong force or strong interaction is needed. While FCC uses electromagnetic forces, SAM uses primarily electrostatic forces for stability.

3 FCC uses protons and neutrons, whereas SAM uses protons and (nuclear) electrons. By replacing neutrons with PEPs, we have introduced a major simplification; neutrons don't show up as independent entities, allowing SAM structures to be discovered much more readily than the FCC equivalents.

4 FCC works with a fixed lattice structure, where nucleons are placed at specific sites. In contrast, the *SAM structure is adaptable* and changes as the number of nucleons increases. However, because nucleons in SAM are not forced into random positions in a *fixed* matrix, they can situate themselves according to a minimum energy configuration.

5 With SAM, we have discovered the specific rules that govern the *growth* trajectories of the elements, consistent with known elemental properties. Growth is based on the ending concept (i.e., the placement of the next proton in the structure is easily determined). Because of the need for carbon nuclets to be completed, this iterative growth process converges rapidly to specific elements and isotopes.

It is not a coincidence that we have been able to construct a largely complete periodic table of elements that has all the earmarks of being in accordance with observed nuclear parameters and properties, including many isotopic configurations. In this discovery process, properties such as stability and radioactive decay almost automatically reveal themselves.

7.2 NUCLEAR STRUCTURE AND QUANTUM MECHANICS

A recent development in nuclear structure theory has been the award of the 2016 Lise Meitner Prize by the European Physical Society (EPS). The prize was awarded to Ulf Meissner for his "Nuclear Lattice Effective Field Theory." For Dr. Norman Cook this was highly significant, because ". . . the explicit rejection of Bohr's philosophy (i.e., the

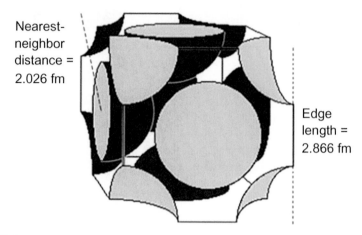

Nearest-neighbor distance = 2.026 fm

Edge length = 2.866 fm

Figure 7.4 Calculation of the nuclear density assuming the close packing of nucleons inside the nuclei. The figure is drawn to scale in which each of the nucleons has a radius of 0.862 fm and a core density of 0.17 nucleons/fm³.
[Adapted from Cook 2010, 127]

Copenhagen interpretation of the uncertainty principle) means that detailed geometrical explanations of nuclear structure are now 'allowed'" [Cook & Di Sia 2018]. Thus, it appears that at least the European Physical Society recognizes that important work is being done in nuclear structure theory.

In his book *Models of the Atomic Nucleus* [Cook 2010], Cook reviews the experimental values of nuclear density and concludes that "nuclei are at least half filled with nucleon matter—in first approximation, a dense liquid or a solid, rather than a diffuse gas."

Figure 7.4 is a depiction by Cook based on FCC close packing. With the indicated dimensions, a density of 46% is calculated, making it obvious that there is not much room inside the nucleus for the nucleons to move around freely.

According to Hodgson [Hodgson et al. 1997, 315] as quoted by Cook:

We know the cross-section for the interaction of two free nucleons, and this gives a mean free path that is far too short to be compatible with independent motion inside the nucleus. We can accept that different models should reflect different aspects of the nucleus, but they should be consistent with one another. [Cook 2010, 94]

Looking at the structure of the nucleus in SAM, we can do a similar density calculation based on the volume of a single proton times the number of protons divided by the MBS volume (minimal bounding sphere volume). However, we recognize that this number does not say much since the minimal bounding sphere is not a good volume approximation for the nucleus. A minimal bounding ellipsoid might be better suited. The MBS-based density varies between 49% (carbon) and 6% (tellurium). With a minimal bounding ellipsoid as the base, the density numbers would be closer to the

FCC numbers. If we look at the backbone of carbon nuclets there is no free path at all to pass through. It is a very rigid structure.

Thus, it is not hard to conclude that quantum mechanics, with the built-in assumptions of the Heisenberg Uncertainty Principle and the Pauli Exclusion Principle, is not applicable to structured nuclei like in SAM. Neither is any calculation based on the independent particle model applicable; the basic assumptions do not apply to SAM.

7.3 STANDARD MODEL UNDERSTANDING OF BINDING ENERGY

We already mentioned the binding energy of a nucleus as a significant factor in determining possible reactions for a given nucleus (Table 2.2; Section 2.17). Nuclear binding energy is the *minimum energy required* to break up a nucleus into its fundamental components. Therefore, this energy is released when the nucleus comes together to create deuterons and structure. This is the reason *higher* binding energy is preferred for the nucleus as it is a *lower* energy state. This seems counterintuitive, but it isn't. If the binding energy goes *up*, energy is being released, so the overall energy level of the nucleus goes *down*. This released energy needs to be spent again if you want to break up the nucleus into its components.

As mentioned before, binding energy (BE) can also be defined as the energy that was *released when the nucleus came together* from its components to create deuterons and adding structure. However, it would then carry a negative sign. Also, we have seen that it is very unlikely that atomic nuclei come together from the smallest components (protons, "neutrons") in a single complete fusion step since a fusion process that uses bigger chunks or a buildup in small incremental steps is more likely.

Typically, the binding energy is averaged per nucleon (proton and neutron) in the nucleus and plotted against the number of nucleons in the nucleus (Fig. 7.5). Data was compiled from International Atomic Energy Agency [IAEA] data, listed in Table F.1, which can be found in Appendix F.

The series of light elements from hydrogen up to sodium is observed to exhibit generally increasing binding energy per nucleon. The region of increasing binding energy is followed by a region of relative stability in the sequence from magnesium through to xenon. Finally, in elements heavier than xenon, there is a decrease in binding energy per nucleon as the atomic number increases. The high binding energy value of helium-4 relative to the surrounding isotopes is of particular interest. Looking at the average binding energy per nucleon we may conclude that iron-56, with a binding energy per nucleon of 8.8 MeV, displays the third most tightly bound nucleus. Its average binding energy per nucleon is exceeded only by iron-58 and nickel-62. The nickel isotope has the most tightly bound nucleus—at least according to the Standard Model.

The peak of the curve is explained as a balance between the Coulomb force and the strong force. As the nucleus gets bigger in larger elements, the strong force is only "felt" over short distances, between neighboring nucleons. The Coulomb force can be "felt" throughout the nucleus, even as it gets bigger. The curve is also used to explain fusionable and fissionable nuclei: lighter nuclei than iron are more likely to release energy in a fusion process, and heavier nuclei than iron are more likely to

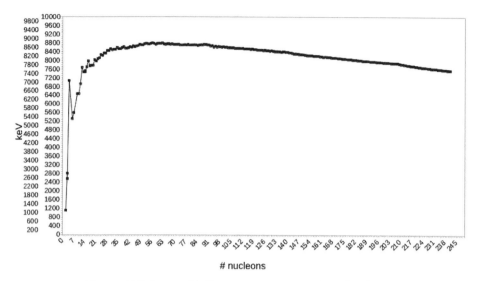

nucleons

Figure 7.5 Average binding energy versus number of nucleons.
[Compiled from IAEA 2021 data]

release energy in a fission process. We should realize that this argument is being made based on the curve of average binding energy per nucleon.

When we used the term binding energy prior to this section, we used it not as average binding energy per nucleon of a nucleus, but as the total binding energy of a nucleus. The reason, of course, was that we were interested in the behavior of the whole nucleus and the attributes of the atom—not in an average number with no real meaning to the problem at hand. Interestingly, the average binding energy (BE) is derived from the total binding energy of a nucleus, but it is hard to find this data or even a chart in the common literature where the total binding energy in megaelectron volts is plotted against the elements/isotopes (number of nucleons). Figure 7.6 was created using the IAEA data mentioned before (Table F.1).

It should be mentioned that the binding energy of the whole atom can also be calculated. This includes the outer electrons. Looking at the numbers, the binding energy provided by the outer electrons is negligible compared with the nucleus. However, can the binding energy of the atom or the nucleus actually be measured? Except for small atomic nuclei, no one has ever broken up a nucleus into its fundamental components (single protons/"neutrons") or assembled one in one step from its fundamental components, creating deuterons and adding structure.

What is being measured is the actual mass of nuclei and it is being compared with the theoretical mass, which is based on the known mass of a proton, "neutron," and electron. The binding energy is equated with the mass defect (the difference between actual and theoretical mass) through $E = mc^2$, the famous mass–energy equivalence formula popularized by Albert Einstein. The missing mass is equal to the released energy when the components came together and formed the nucleus. Mass of atoms or nuclei is often presented in atomic mass units (amu), which is a fixed value equal to one-twelfth of the mass of an unbound atom of carbon-12.

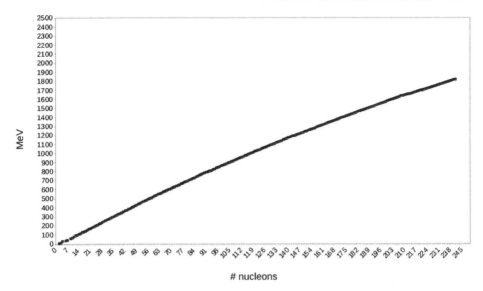

Figure 7.6 Total binding energy versus number of nucleons.
[Compiled from IAEA 2021 data]

There exists a so-called semi-empirical mass formula (SEMF), based on the liquid drop model, which is being used to estimate the actual mass of a nucleus. As the name suggests, it is based partly on theory and partly on empirical measurements. It was first formulated in 1935 by Carl Friedrich von Weizsäcker and although refinements have been made to the coefficients over the years, the structure of the formula remains the same today. The formula gives a good approximation for atomic masses. However, it fails to explain the existence of values of greater binding energy at certain numbers of protons and "neutrons" (Fig. 7.6)

The corresponding mass formula is defined purely in terms of the numbers of protons and "neutrons" making up the isotope. The original Weizsäcker formula defines five terms:

- volume energy,
- surface energy,
- Coulomb energy,
- asymmetry energy, and
- pairing energy.

The coefficients included in the formula are calculated by fitting to experimentally measured masses of nuclei. The semi-empirical mass formula provides a good fit to heavier nuclei and a poor fit to very light nuclei, especially helium-4. To explain those peaks, another model—the shell model of the nucleus—is being used. The highest average binding energy value per nucleon is reached—according to the formula—with copper-63, not nickel-62, but it is close.

When we look at actual mass data of atomic nuclei we have to be aware that sometimes values are based on experiments and sometimes not, being theoretical calculations with the help of SEMF. SEMF is based on the liquid drop model. However, there are other variations of SEMF that are based upon other models.

What we can conclude for now is:

1 Nuclear binding energy revolves around the definition of the mass defect.
2 The available data is sometimes questionable—it is not always based on experiments and is sometimes calculated using variations of SEMF.
3 Defining nuclear binding energy and mass defect to be equal through $E = mc^2$ is an assumption.

Our first question must then be: What is mass? The textbook answer is:

Mass, in physics, is a quantitative measure of inertia, a fundamental property of all matter. It is, in effect, the resistance that a body of matter offers to a change in its speed or position upon the application of a force. The greater the mass of a body, the smaller the change produced by an applied force. [Encyclopedia Britannica 2021/science/mass-physics]

The question is now: Is the *mass defect* real, does it tell the full story? Let's look at an example, a very typical $\beta-$ decay (Tables 7.2–7.5), going from neon-23 to sodium-23, where:

Proton:	1.007276466879 amu
Neutron:	1.00866491588 amu
Electron:	0.000548579909070 amu
Neutron–proton:	0.001388449 amu
Conversion factor:	931.44 MeV/amu (based on carbon-12)
Neutron self-binding energy:	0.781906622 MeV
$\beta-$ decay:	0.004697618 MeV

First, we calculate the theoretical masses and convert them into energies (Table 7.2). Then we look at the actual masses and convert them into energies too (Table 7.3). After that we look at the mass as well as the energy differences (Table 7.4). As a last step we compare the differences (Table 7.5).

Table 7.2 Theoretical mass and energy based on the component values of Ne-23/Na-23.

	Protons 1.007276466879 amu	"Neutrons" 1.00866491588 amu	Electrons 0.000548579909070 amu	Mass (amu)	Energy (MeV) using conversion factor
Ne-23	10	13	10	23.190894374	21600.926656017
Na-23	11	12	11	23.190054505	21600.14436835

Table 7.3 Actual mass of Ne-23/Na-23.

Isotope	Mass (amu)	Energy (MeV)
Ne-23	22.9944669	21417.966249336
Na-23	22.98976928199	21413.590700017
β– decay	0.004697618	4.375549319

Table 7.4 Theoretical–actual (T–A) mass differences for Ne-23/Na-23.

Isotope	Theoretical mass (amu)	Actual mass (amu)	T–A difference (amu)	T–A difference (MeV)
Ne-23	23.190894374	22.9944669	0.196427474	182.960406383
Na-23	23.190054505	22.98976928199	0.200285223	186.55366812
β– decay		0.004697618		
Ne-23–Na-23	0.000839869	0.004697618	**−0.003857749**	−3.593261737
Difference MeV	**0.782287581**	**4.375549319**	−3.593261729	

Table 7.5 Comparing the differences for Ne-23/Na-23.

Isotope	Actual mass (amu)	T–A difference (amu)	T–A difference (MeV)
Ne-23–Na-23	0.004697618	−0.003857749	−3.593261737
β– decay	0.004697618		
Difference	0.000000000		
Actual mass difference (MeV)	**0.000000000**		

If we look at the actual masses and the mass of a β– decay in the example above the difference is negligible. This does not leave much room for any more participants. This is not surprising as the β– decay energy is usually calculated from the actual mass difference. Looking at the theoretical masses there is however a difference (Table 7.6).

The mass difference between a neutron and a proton (0.001388449 amu) and an electron (0.000548579909070 amu) amounts to 0.000839869 amu, which is equal to

Table 7.6 Theoretical mass differences of a proton and a "neutron."

	Theoretical mass (amu)	Energy (MeV)
Neutron/PEP	1.00866491588	939.510849247
Proton	1.007276466879	938.21759231
Electron	0.000548579909070	0.510969271
Neutron–proton–electron	0.000839869	**0.782287666**
		0.782287581

0.782287666 MeV or 0.782287581 MeV through another calculation path (different rounding):

0.782287581 MeV – –3.593261737 MeV = 4.375549318 MeV.

The decay of the PEP supplies 0.782287581 MeV. With this change we gain 3.593261737 MeV binding energy. The sum of both energies is radiated as β– decay (4.375549318 MeV).

It appears that an electron bound to a proton in the nucleus as part of a PEP has more mass than an electron outside the nucleus (Table 7.6). This is actually *negative binding energy*, this energy amount of 0.782287581 MeV needs to be spent to put an electron and a proton together as a PEP. It is not released because of the gain in mass we see. When the PEP splits into a proton and an electron, this energy will become available again as we see above. It contributes to the overall amount of energy that needs to be radiated away.

Next, we look at a special case of β– decay, one in which the binding energy actually drops and hydrogen-3 decays to helium-3 (Tables 7.7–7.10), with:

Proton:	1.007276466879 amu
Neutron:	1.00866491588 amu
Electron:	0.000548579909070 amu
Neutron–proton:	0.001388449 amu
Conversion factor:	931.44 MeV/amu (based on carbon-12)
Neutron self-binding energy:	0.781906622 MeV
β– decay:	0.018589493 MeV

Table 7.7 Theoretical mass and energy based on the component values of H-3/He-3.

	Protons 1.007276466879 amu	"Neutrons" 1.00866491588 amu	Electron 0.000548579909070 amu	Mass (amu)	Energy (MeV)
H-3	1	2	1	3.025154879	2817.750260075
He-3	2	1	2	3.024315009	2816.967972408

Table 7.8 Actual mass of H-3/He-3.

Isotope	Mass (amu)	Energy (MeV)
Hydrogen-3	3.0160492779	2809.268939407
Helium-3	3.0160293201	2809.250349914
β– decay	0.0000199578	0.018589493

Table 7.9 Mass differences for H-3/He-3.

Isotope	Theoretical mass (amu)	Actual mass (amu)	T–A difference (amu)	T–A difference (MeV)
Hydrogen-3	3.025154879	3.0160492779	0.009105601	8.481321089
Helium-3	3.024315009	3.0160293201	0.008285689	7.717622069
β– decay		0.0000199578		
H-3–He-3	0.00083987	0.000019958	0.000819912	**0.763698833**
Difference MeV	**0.782288513**	**0.01858968**	**0.763698833**	

Table 7.10 Comparing the differences for H-3/He-3.

Isotope	Actual mass (amu)	T–A difference (amu)	Difference (MeV)
H-3–He-3	0.000019958	0.000819912	0.763698833
β– decay	0.0000199578		
Difference	0.0000000002		**0.763698833**
Actual mass difference (MeV)	0.000000186		

Interestingly, the binding energy difference between hydrogen-3 and helium-3 is 0.763698833 MeV. The difference between the prior difference and the binding energy difference is equal to 0.01858968 MeV, which is the energy of the β– decay in this case:

0.782288513 MeV – 0.01858968 MeV = 0.763698833 MeV.

The β– decay releases enough energy to cover the binding energy that must be spent to reach helium-3 from hydrogen-3, as well as a small amount of energy which we see as β– decay energy. A quick check with our list of β– decays (Section 2.17.1) shows, that in every case we listed where the binding energy dropped during the β– decay, it was in a range of less than 0.781906622 MeV. This means that in every β– case we have seen so far which involves decreasing binding energy, that difference is covered by the energy released through the breakup of the PEP. Otherwise such decay steps would not be possible. β– decay can happen if the binding energy goes down, but only if the drop in binding energy is less than the energy the breakup of the PEP provides. We are confident this pattern will hold up.

Now we look at a typical β+ decay/e-capture from Na-22 to Ne-22 (Tables 7.11–7.13), with:

Proton:	1.007276466879 amu
Neutron:	1.00866491588 amu
Electron:	0.000548579909070 amu
Neutron–proton:	0.001388449 amu
Conversion factor:	931.44 MeV/amu (based on carbon-12)
Neutron self-binding energy:	0.781906622 MeV

Table 7.11 Theoretical mass and energy based on the component values of Na-22/Ne-22.

	Protons 1.007276466879 amu	"Neutrons" 1.00866491588 amu	Electrons 0.000548579909070 amu	Mass (amu)	Energy (MeV)
Na-22	11	11	11	22.181389589	20660.633519103
Ne-22	10	12	10	22.182229458	20661.41580677

Table 7.12 Actual mass of Na-22/Ne-22.

Isotope	Mass (amu)	Energy (MeV)
Na-22	21.994437418	20486.498788622
Ne-22	21.991385109	20483.655745927

Table 7.13 Mass differences of Na-22/Ne-22.

Isotope	Theoretical mass (amu)	Actual mass (amu)	T–A difference (amu)	T–A difference (MeV)
Na-22	22.181389589	21.994437418	0.186952171	174.134730156
Ne-22	22.182229458	21.991385109	0.190844349	177.760060433
Na-22–Ne-22	−0.000839869	0.003052309	−0.003892178	−3.625330276
Difference MeV	−0.782287581	2.843042695	−3.625330276	

In the β+ decay step an electron enters the nucleus and converts a proton to a PEP. The first step involves resetting the structure of the nucleus. With this change

3.625330276 MeV of binding energy is gained. However, a single proton is now located where a PEP should be. The creation of a PEP by taking in an electron requires 0.782287581 MeV, as discussed earlier. The sum of both energies is radiated away by the nucleus (2.843042695 MeV). How this happens is currently of no concern to us. It is evident from the published data that this amount of energy being radiated away is the amount of energy being seen in total as described earlier:

−0.782287581 MeV − −3.625330276 MeV = 2.843042695 MeV.

Here is another typical β+ decay/e-capture from K-40 to Ar-40 (Tables 7.14–7.16), with:

Proton:	1.007276466879 amu
Neutron:	1.00866491588 amu
Electron:	0.000548579909070 amu
Neutron–proton:	0.001388449 amu
Conversion factor:	931.44 MeV/amu (based on carbon-12)
Neutron self-binding energy:	0.781906622 MeV

Table 7.14 Theoretical mass and energy based on the component values of K-40/Ar-40.

	Protons 1.007276466879 amu	"Neutrons" 1.00866491588 amu	Electrons 0.000548579909070 amu	Mass (amu)	Energy (MeV)
K-40	19	21	19	40.330639122	37565.570504218
Ar-40	18	22	18	40.331478992	37566.352791885

Table 7.15 Actual mass of K-40/Ar-40.

Isotope	Mass (amu)	Energy (MeV)
K-40	39.963998166	37224.06645173904
Ar-40	39.96238312378	37222.562136813643

Table 7.16 Mass differences for K-40/Ar-40.

Isotope	Theoretical mass (amu)	Actual mass (amu)	T–A difference (amu)	T–A difference (MeV)
K-40	40.330639122	39.963998166	0.366640956	341.504052057
Ar-40	40.331478992	39.96238312378	0.369095868	343.790655495
K-40–Ar-40	−0.00083987	0.001615042	−0.002454912	−2.286603438
Difference MeV	−0.782288513	1.504314925	−2.286603233	

In this case, after the energy to create the PEP was used, only 1.504314925 MeV of the original 2.286603233 MeV binding energy difference is left over to be radiated.

The following example is a low-energy β+ decay/e-capture from Ca-41 to K-41 (Tables 7.17–7.19), with:

Proton:	1.007276466879 amu
Neutron:	1.00866491588 amu
Electron:	0.000548579909070 amu
Neutron–proton:	0.001388449 amu
Conversion factor:	931.44 MeV/amu (based on carbon-12)
Neutron self-binding energy:	0.781906622 MeV

Table 7.17 Theoretical mass and energy based on the component values of Ca-41/K-41.

	Protons 1.007276466879 amu	"Neutrons" 1.00866491588 amu	Electrons 0.000548579909070 amu	Mass (amu)	Energy (MeV)
Ca-41	20	21	20	41.338464169	38504.299065798
K-41	19	22	19	41.339304038	38505.081353465

Table 7.18 Actual mass of Ca-41/K-41.

Isotope	Mass (amu)	Energy (MeV)
Ca-41	40.962277921	38153.904146736
K-41	40.96182525796	38153.482518274

Table 7.19 Mass differences for Ca-41/K-41.

Isotope	Theoretical mass (amu)	Actual mass (amu)	T–A difference (amu)	T–A difference (MeV)
Ca-41	41.338464169	40.962277921	0.376186248	350.394918837
K-41	41.339304038	40.96182525796	0.37747878	351.59883488
Ca-41–K-41	−0.000839869	0.000452663	−0.001292532	−1.203916043
Difference MeV	−0.782287581	0.421628462	−1.203916006	

In this case the energy difference is even smaller and only 0.421628462 MeV remains to be radiated.

Now we will have a look at a typical α decay (Tables 7.20–7.22). Radon-215 (Fig. 7.7) is unstable and emits an α particle after 2.30 μs, creating polonium-211 (Fig. 7.8) in the process. According to the isotope tables nothing else is emitted:

Proton:	1.007276466879 amu
Neutron:	1.00866491588 amu
Conversion factor:	931.44 MeV/amu (based on carbon-12)

Table 7.20 Theoretical mass and energy based on the component values for Rd-215/Po-211.

	Protons 1.007276466879 amu	"Neutrons" 1.00866491588 amu	Electrons 0.000548579909070 amu	Mass (amu)	Energy (MeV)
Rd-215	86	129	86	216.790728172	201927.555848802
Po-211	84	127	84	212.757748247	198171.077027146
α	2	2	0	4.031882766	3755.456883114
He-4	2	2	2	4.032979925	3756.478821655

Table 7.21 Actual mass for Rd-215/Po-211.

Isotope	Mass (amu)	Energy (MeV)
Rd-215	214.998745	200258.4310428
Po-211	210.9866532	196521.408256608
α	4.001506179127	3727.162915486
He-4	4.00260325413	3728.184775027

Table 7.22 Mass differences for Rd-215/Po-211.

	Theoretical mass (amu)	Actual mass (amu)	Difference (amu)	Difference (MeV)
Rd-215	216.790728172	214.998745	1.791983172	1669.124805728
Po-211	212.757748247	210.9866532	1.771095047	1649.668770578
α	4.031882766	4.001506179127	0.030376587	28.293968077
He-4	4.032979925	4.00260325413	0.030376671	28.294046315
Rn-215–Po-211α	0.001097159	0.010585621	−0.009488462	−8.837932927
Difference MeV	1.021937779	9.859870824	−8.837933045	
Rn-215–Po-211–He-4	0.00	0.009488546	−0.009488546	−8.838011165
Difference MeV	0.00	8.838011165	−8.838011286	

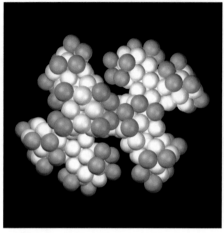
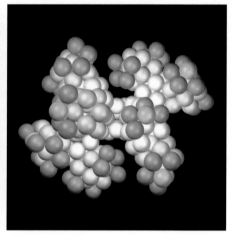

Figure 7.7 Radon-215 before α decay.   **Figure 7.8:** Polonium-211 after α decay from radon-215.

The difference in theoretical mass in the α calculation represents the two electrons the balance sheet is off by. The two electrons will probably be pulled from the electron cloud, creating a real helium-4 atom. The helium-4 lines in the tables show this.

When creating a radon-215 nucleus, binding energy amounting to 1669.124805728 MeV was in theory released. Radon-215 is unstable and immediately emits an α particle. In order to do so, and creating polonium-211 in the process, at least 19.45603515 MeV needs to be spent to meet the binding energy requirement (1649.668770578 MeV) of polonium-211.

Where is this energy coming from after 2.30 μs? The only source is the creation of the α particle. We gain 23.844 MeV from combining two deuterons (4.45 MeV already hidden in the existing deuterons; 28.294 MeV–2.225 MeV–2.225 MeV = 23.844 MeV) in this step (Figs. 7.7 and 7.8). However, 8.838011165 MeV is not accounted for. This is made available as the kinetic energy of the α particle, partially also as the recoil of polonium-211.

We used binding energy in Part I without knowing what it really was when developing the elements in our scheme (Table 2.2, Section 2.17), explaining with it the half-life of some isotopes. We now have a better understanding of what is going on.

The difference between the binding energy of the target and the binding energy of the source of a nuclear reaction is key. If the difference is negative, it needs to be either put into the system or sourced from it by the reaction itself (e.g., creating an α particle). If it is positive, the amount that is not used up within the structure (e.g., creating a PEP) needs to be radiated. If the energy requirements cannot be balanced then the reaction cannot happen if no other energy source is available.

Let us return to the question raised on page 118, that is, is the mass defect real and does it tell the whole story? Our answer is that it seems so, there is apparently no room for additional particles with mass, all energy is accounted for.

## CHAPTER 8

Elemental forces and energies in the nucleus

As described before, other models of the nucleus use at least three forces:

- Strong force (for binding the nucleons (protons and neutrons) together)
- Weak force (for radioactive decay)
- Electrostatic/magnetic force.

The invention of the strong force is the result of the invention of the "neutron." With only positively charged and neutral nucleons placed in the nucleus, the electrostatic repulsion of the protons needed to be overcome. Why was the "neutron" even suggested? The answer is very simple: the proton–electron hypothesis of the nucleus was found to be incompatible with emerging quantum mechanics and Heisenberg's Uncertainty Principle in particular. This is why the old model was abandoned. If you read between the lines it is as simple as that. We will address the weak force in Section 11.1.

8.1 ATOMIC AND NUCLEON RADII

The root mean square (rms) charge radius is one measure of the size of an atomic nucleus, particularly proton distribution. It can be measured by the scattering of electrons by the nucleus. Relative changes in the rms nuclear charge distribution can be measured accurately using atomic spectroscopy. This definition of charge radius can also be applied to a proton and a PEP.

Direct measurements are based on precision measurements of the atomic energy levels in hydrogen and deuterium and measurements of scattering of electrons by nuclei. There is much interest in knowing the charge radii of protons and deuterons, as these can be compared with the spectrum of atomic hydrogen/deuterium.

The 2018 CODATA-recommended values are:

Proton, $R_p = 0.8414(19) \times 10^{-15}$ m [NIST 2018/proton + radius]
Deuteron, $R_d = 2.12799(74) \times 10^{-15}$ m [NIST 2018/deuteron + radius]

Looking at the recommended values over the last few years, the proton radius is shrinking as measuring methods become better.

The "classical" radius of an electron has been calculated at 2.82×10^{-15} m, which would make it 3.4 times bigger than the proton. The classical electron radius is built from the electron mass, the speed of light, and the electron charge [Wikipedia 2021/ Classical_electron_radius]. However, currently there is different thinking:

> *According to modern understanding, the electron is a point particle with a point charge and no spatial extent. Attempts to model the electron as a non-point particle have been described as ill-conceived and counter-pedagogic.* [Wikipedia 2021/ Classical_electron_radius]

This quote embodies one of the biggest fallacies of modern physics. A *point* is a *mathematical concept*. You can use it to describe a physical object, you can use it to simplify calculations—once you are sure that this simplification does not oversimplify. However, you can never *substitute* the physical object with a concept as done here. You are not allowed to objectify a concept. Unfortunately, this has become the norm in nuclear- and astrophysics over the last 120 years.

8.2 ELECTRIC FORCES AND POTENTIAL ENERGY

Wikipedia defines "electric potential energy" and "electrostatic potential energy" as:

> *Electric potential energy is a potential energy (measured in joules) that results from conservative Coulomb forces and is associated with the configuration of a particular set of point charges within a defined system. An object may have electric potential energy by virtue of two key elements: its own electric charge and its relative position to other electrically charged objects. The term 'electric potential energy' is used to describe the potential energy in systems with time-variant electric fields, while the term 'electrostatic potential energy' is used to describe the potential energy in systems with time-invariant electric fields.*
> . . .
> *The electric potential energy of a system of point charges is defined as the work required to assemble this system of charges by bringing them close together, as in the system from an infinite distance. Alternatively, the electric potential energy of any given charge or system of charges is termed as the total work done by an external agent in bringing the charge or the system of charges from infinity to the present configuration without undergoing any acceleration.*
> . . .
> *The electrostatic potential energy, U_E, of one point charge q at position r in the presence of an electric potential is defined as the product of the charge and the electric potential.* [Wikipedia 2021/https://en.wikipedia.org/wiki/ Electric_potential_energy]

The derivation of U_E is based on the Coulomb force law. However, the law is based on point charges and it looks at the force between charges without considering the shadowing effects of other charges present. If the charge radius of the electron (not necessarily the electron itself) is indeed bigger than the proton, then this further complicates the calculation. Calculating values under these circumstances makes no sense. The most we can do and if we accept the simplification to point charges for a moment, we can use the Coulomb force law to show, that in a simple configuration attraction is higher than repulsion. We start with the Coulomb force law:

$$F = k_e \frac{q_1 q_2}{r^2}.$$

The force between two protons with radius r and distance $2r$ would then be:

$$F_{pp} = k_e \frac{qq}{(2r)^2} = k_e \frac{qq}{2^2 r^2}.$$

The force between a proton and an electron with distance r would then be:

$$F_{pe} = k_e \frac{-qq}{r^2}.$$

The factor of $-1/4$ $(-1/2^2)$ is easily visible between the last two formulas. Thus, the repelling proton–proton force is one-quarter of the attractive proton–electron force given the negative charge positioned between the positive charges.

8.3 MORE THOUGHTS ON OUTER ELECTRONS

The same train of thought, of course, applies to a chain of inner electron, proton, and a possible outer electron if it ever comes close enough to the nucleus. If the outer electrons are removed through ionization this situation might become possible. However, such an electron attaching to the outside of the nucleus would then only have one anchor point—not the required two. It cannot stay there. An attachment like this can only happen through a PEP where the proton of the PEP provides the second anchor point.

An outer electron will always be attracted by the overall positive nucleus of an atom. At the same time it is repelled by all the other outer electrons of the atom. Those outer electrons create a shield around the nucleus. Only if there is room inside a nucleus to take in an electron and the step is energetically viable will this happen ($\beta+$ decay).

In the case of hydrogen-1 there are no other outer electrons and there is no inner electron. Why does the outer electron stay away from the nucleus? In Section 7.3 we learned that energy is needed to create a PEP from a proton and an electron, 0.782287581 MeV to be precise. If this energy is not available then it will not happen, despite the Coulomb force.

8.4 POTENTIAL ENERGY VERSUS BINDING ENERGY

There might be a correlation between the mass of the nucleus, its energetic equivalent, and the calculation of the potential energy of a nucleus as described above, most likely in a much more complicated way than laid out before. The binding energy reduces this potential energy of that system by a given amount, based on the structure. Despite potential energy as well as binding energy being based on the structure of a nucleus, it is in our opinion too early to assume a correlation exists between them. This topic needs further research.

For the potential energy alone we can assume that it at least partially contributes to the mass of a particle even in a single-particle system—the charge of the particle creates a field that stores energy.

## CHAPTER 9

Binding energy and mass defect in SAM

In Section 7.3 we recognized that the semi-empirical mass formula (SEMF) was based on the liquid drop model. Assuming that the mass defect has something to do with the binding energy, and having a model that has a nucleus with a structure like SAM, we would expect an impact on the calculation of binding energies in SAM. We can say with certainty that SEMF is unusable in the context of SAM.

How would the binding energy of a nucleus be calculated in SAM? How would the structure determine the values? We found that the organizational patterns (single nuclet, two nuclets, branching, elongation (Section 3.6)) will affect the binding energy numbers.

9.1 FIRST-ORDER ORGANIZATIONAL PATTERN

There are two lines of thought to be considered. (1) The simplest nucleus with protons bound together is the deuteron which consists of two protons and an inner electron "between" the protons. We will accept the binding energy value of 2.225 MeV for the deuteron as it is the simplest case. We also recognize the important role of the deuteron as an integral piece in constructing the endings that make up the nuclei. We call the deuteron a "line connection" or a "line" and we will determine the number of "lines" for the first nuclei (Table 9.1, see "Remarks" column). (2) The mass calculation in the Standard Model is based upon carbon. With SAM, we will do the same. This makes even more sense when looking at the 2nd and 3rd organizational patterns. Those patterns are related to the creation of the backbone of the nucleus, which is built up from carbon nuclets (Table 9.2).

The special cases for hydrogen-3 and helium-4 can be made plausible by looking at their configurations regarding possible overlapping inner electron spheres of influence (Figs. 9.1 and 9.2).

9.2 SECOND-ORDER ORGANIZATIONAL PATTERN

This pattern, although strictly only valid for the creation of the second nuclet, will be used as a template for all other nuclets. However, we only take the existence of the

Table 9.1 SAM lines and binding energy for initial isotopes.

Nucleus	# Lines	SAM line BE (MeV)	Remarks
Hydrogen-2	1	2.225	By definition
Hydrogen-3	2 × 2 = 4	8.900	The two inner electrons are overlapping, doubling the number of lines (Fig. 9.1)
Helium-3	3	6.675	One inner electron, no overlapping
Helium-4	6 × 2 = 12	26.700	The two inner electrons are overlapping, doubling the six lines a helium-4 nucleus has (Fig. 9.2)
Lithium-6	15	33.375	12 lines + 6 = 18 for the new deuteron in this case, but one line is now broken, produced by 3 electrons covering it, so we must deduct 3, leaving 15
Lithium-7	19	42.275	The new proton–electron pair (PEP) adds four more lines
Beryllium-9	27	60.075	A deuteron counts as eight lines now, because there are enough protons available, 19 + 8
Boron-10	31	68.975	The new proton–electron pair adds four more lines
Boron-11	34	75.650	The new proton–electron pair adds three more lines
Carbon-12	41	91.225	+5 by algorithm = 39, but by definition 41 = closest to 92.164 MeV, which is the known value of BE

nuclets/endings into account—not their positions. We also define a new short code to describe the nuclet/ending (Table 9.2).

9.3 THIRD-ORDER AND FOURTH-ORDER ORGANIZATIONAL PATTERN

For our calculations we only add the lines of carbon nuclets and endings for each nucleus, we do not consider the positioning of the endings to each other or even the positioning of the branches relative to each other (3rd order) as well as the elongation (4th order) of the nucleus. If indeed this line approach makes sense, we would for this reason expect it to break down once branching happens in a recognizable

Table 9.2 SAM lines per nuclet/ending.

Nuclet/ending	SAM # lines	Short code	Remarks
Prime carbon	41	p	See above
Deuteron (two-ending)	8	2	Four connections per proton
Four-ending	16	4	Four connections per proton
Five-ending	21	5	The additional proton–electron pair makes five more connections
Lithium nuclet	24	h	Four connections for one additional PEP based on the five-ending minus one for an opened-up line (additional space); $4 - 1 = 3$
Beryllium ending	32	e	Add another deuteron = eight
Boron ending	40	b	Add another deuteron = eight
Carbon nuclet	45	c	Add another proton to reach the icosahedron structure (40), which equals five connections
Additional PEP with three connections	3	3	Three additional connections in a tetrahedron setup

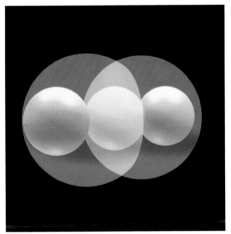

Figure 9.1 Hydrogen-3 with inner electrons twice the size of a proton.

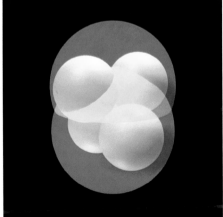

Figure 9.2 Helium-4 with overlapping inner electrons.

manner—which means beyond iron we would expect to see a breakdown of the correlation in one direction, even if we actually have a good correlation for nuclei between carbon and iron.

9.4 LINE CALCULATION

As an example we look at the structure of scandium-45 and determine the components and their given lines using the short code defined above: $p + c + c + h + 5 = 41 + 45 + 45 + 24 + 21 = 176$. Each element and isotope has therefore a specific number of "lines" and a simple multiplication by 2.225 MeV gives us the total binding energy (BE) for the elements (and isotopes) in SAM. Using the values in Table 9.1, as well as the line calculation formula (Table 9.2), SAM values are calculated and compared with known IAEA binding energy data (Table 23.1). First, we create a graph for average SAM binding energy versus number of protons (Fig. 9.3), then a total SAM binding energy graph (Fig. 9.4). We see that the SAM values do not drop after nickel. This is a clear, but not unexpected, deviation from the literature values. We emphasized from the start that our SAM line calculation would not take into account branching or elongation of the nucleus.

Let us next overlay the graphs. We start with both average binding energy graphs (Fig. 9.5). Next, we compare both total binding energy graphs (Fig. 9.6). The two curves appear to correlate until the iron/nickel/zinc number range (54–64), but then we see a clear difference between the literature and SAM values. The SAM values in the number range beyond zinc are always higher. If we only look at the basic configuration of the elements (not all the isotopes in between) and plot the change in BE from one element to the next for SAM, as well as the literature values, we see that the

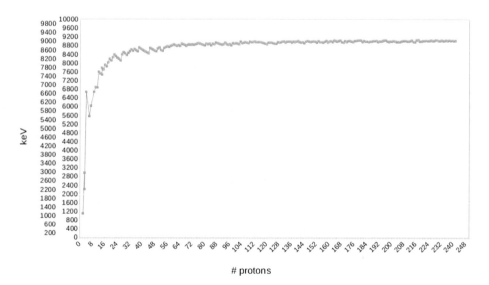

Figure 9.3 Average SAM BE versus number of protons.

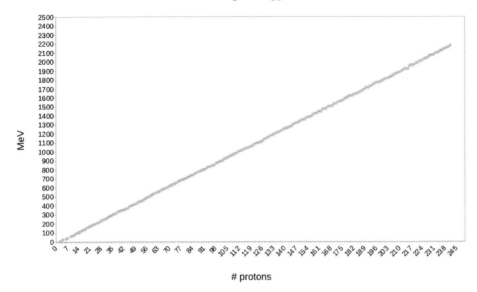

Figure 9.4 Total SAM binding energy versus number of protons.

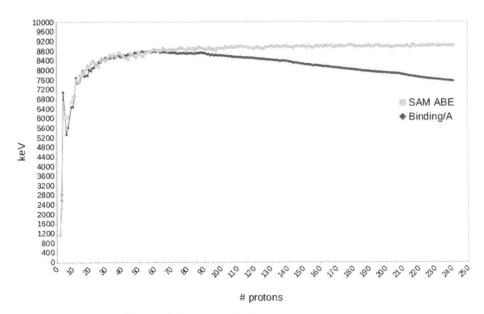

Figure 9.5 Average binding energy comparison.

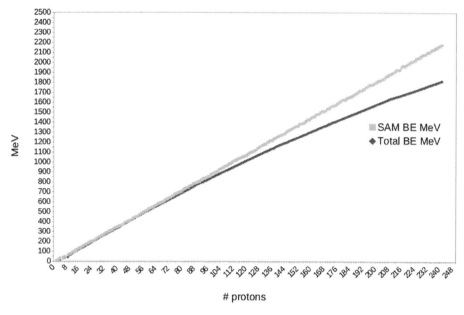

Figure 9.6 Total binding energy comparison.

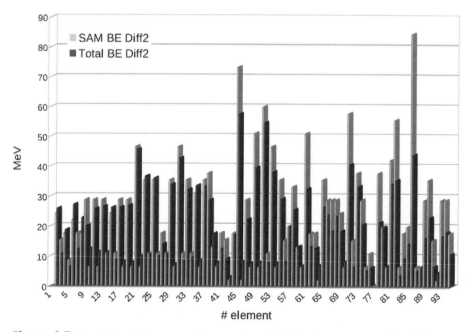

Figure 9.7 Growth deviation comparison per element. Adding two protons and/or PEPs sometimes yields values of around 40 MeV and sometimes as little as 5 MeV.

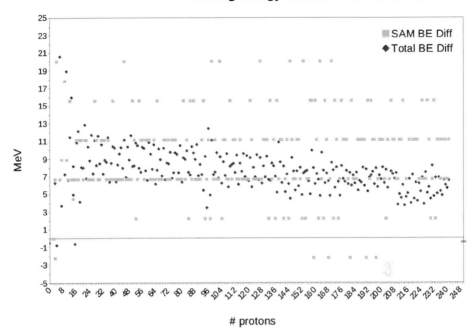

Figure 9.8 Growth differences, SAM BE versus "measured" BE.

correlation even holds true beyond the iron/nickel/zinc area. The changes are kept in sync even though the SAM values get higher. When looking at the binding energy differences stepping from one element to the next, for either literature or SAM values (Fig. 9.7, based on data from Table F.1), we can see that adding two protons and/or PEPs sometimes yields values of around 40 MeV and sometimes as little as a 5-MeV difference in BE. It is obvious that such changes in overall binding energy are the result of structural (re)settling. Otherwise, we would have to accept that the amount of binding energy for the last two added protons really has that much or that little binding energy. Obviously, averaging out these values in typical presentations obfuscates the actual binding energy value for the last two protons added.

We complete the overall picture with two more graphs (Figs. 9.8 and 9.9). Figure 9.8 compares growth patterns. Figure 9.9 shows the error graph for the binding energy difference (%) between actual BE and SAM BE plotted against number of protons.

The ongoing branching of the nucleus opens space between branches. At the larger level, beyond the nuclets we see that the densest packing is not optimal. The nucleus could in theory be more densely packed in a chaotic way or in a lattice structure. Instead, it has a branched structure. This structure causes the added protons to the nucleus above iron not to be in the densest packed state. This diminishes the real binding energy value. The deviation we see taking place happens when the structure of the nucleus develops branches of noticeable size around the elements iron, nickel, and zinc.

Limiting the SAM line calculation to the 1st and 2nd organizational pattern over-states the possible binding energy to be released when the nucleus comes together,

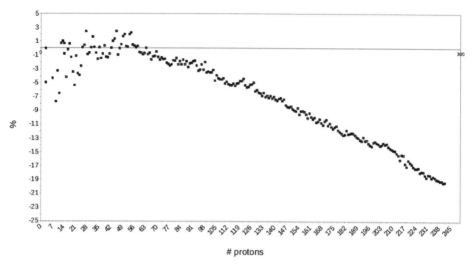

Figure 9.9 Error between "measured" BE and SAM BE.

creating deuterons and adding structure. Instead, the branches of a nucleus "see" each other through the quasi-inner electrons between them. They "stress" the structure and reduce the binding energy to be released when the nucleus comes together. The growth of the nucleus beyond iron is only possible because of a trade-off of stability and densest packing against size.

For now, we take away from this chapter that we have found another very interesting correlation between experimental values and consequences of the model. When there is a deviation we can explain that deviation.

9.5 STRESS ENERGY

The energy that is stored in the structure, because it cannot be released when the isotope is created, needs to be looked at in more detail. This energy we named "stress energy" because the effects of the 3rd-order and 4th-order organizational patterns (especially branching, but also elongation) would diminish the amount of binding energy created through the 1st and 2nd organizational patterns (initial nuclet and further nuclets). Since it diminishes the binding energy the sign of this "stress energy" needs to be negative. This is energy that should have been released simply because of the structure and the effects of spherically dense packing per nuclet. However, if we look at the overall structure of the nucleus, once it gets elongated and branches are created, a different story emerges. It should be noted that the branches with quasi-inner electrons between them stress the overall structure and prevent the full amount of binding energy from being released.

The amount of stress energy is represented by the difference between the *dark-gray line* (Total BE) and the *light-gray line* (SAM BE) in Figure 9.6 (vertical axis) for a given isotope (horizontal axis). The amount of stress energy per nucleus goes up to around –350 MeV with the heavier nuclei (Table F.1). Imagine what we could do with this amount of energy if we were able to release it. We will talk about one way to do this (at least partially) when we discuss the fission process in Chapter 13.

9.6 SAM SEMI-EMPIRICAL BINDING ENERGY FORMULA

In order to get closer to the real binding energy the calculation needs to be based on proton and electron surface and volume charges. Another option would be one or more correction terms for the initial line calculation value to compensate for branching, elongation, and general positioning of nuclets/endings on nuclei. This would be similar to the approach of the Semi-Empirical Mass Formula of the Standard Model, resulting in a Semi-Empirical Binding Energy Formula (SEBEF). This is another research topic.

9.7 CAN SAM LINE BINDING ENERGY HELP WITH THE FLUORINE-18 SITUATION?

When we steal a PEP from fluorine-19 we end up with fluorine-18, There are two natural options to steal a PEP: from the five-ending (A, Fig. 9.10 *left*) and from the two-ending (B, Fig. 9.10 *right*). Which one is the correct one?

Fluorine-18A and fluorine-18B have the same SAM line BE when we handle the PEP and the proton the same way. So, the short answer is: no, the SAM line BE can't help us here. However, configuration A has a slightly smaller minimal binding sphere (MBS) (Table 9.3).

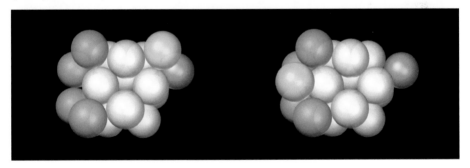

Figure 9.10 Fluorine-18A (*left*) and fluorine-18B (*right*). The remaining PEP is shown in *yellow* and the proton in *brown*.

Table 9.3 Fluorine-18 data.

Isotope	MBS	MBS vol./p#	SAM line BE (MeV)
Fluorine-18A	4.02432	15.1668	153.525
Fluorine-18B	4.06499	15.6313	153.525

Because they are so similar, we would therefore assume that both configurations for fluorine-18 are valid and both could exist. Configuration A might be preferred over B because of the smaller MBS.

Fluorine-18 is unstable and decays with a half-life of 109.739 min to oxygen-18. Because there are two possible versions of fluorine-18, we would expect to see two different decay paths. We will look at the relation of decay paths and structure in more detail in Section 11.13.

CHAPTER 10

A macroscopic view

10.1 ELEMENT ABUNDANCES

We have talked about and used the relative abundance of isotopes of an element, but what about the relative abundance between elements on Earth, in the Sun, and throughout the Universe (as far as we can determine those numbers)? For our Milky Way the mass distribution is as provided in Table 10.1.

Surface of the Sun abundances (Table 10.2) are similar to those of the Milky Way.

Table 10.1 Element abundance in the Milky Way. [Wikipedia 2021/ Abundance_of_the_chemical_elements]

Element	%
Hydrogen	73.90
Helium	24.00
Oxygen	1.040
Carbon	0.460
Neon	0.134
Iron	0.109
Nitrogen	0.096
Silicon	0.065
Magnesium	0.058
Sulfur	0.044

Table 10.2 Element abundance on the Sun's surface. [ThoughtCo 2021/ element-composition-of-sun-607581]

Element	%
Hydrogen	71.00
Helium	27.10
Oxygen	0.970
Carbon	0.400
Silicon	0.099
Nitrogen	0.096
Magnesium	0.076
Neon	0.058
Sulfur	0.040
Iron	0.014

Earth's continental crust element abundance is provided in Table 10.3. The element abundance in the Earth's mantle is provided in Table 10.4.

Note that element abundance in the crust is very different from the values of our stellar and galactic environment and even from the Earth's mantle. Considering this

Table 10.3 Element abundance in the Earth's crust. [PTE 2021/Properties/A/CrustAbundance.an.html]

Element	%
Oxygen	46.00
Silicon	27.00
Aluminum	8.10
Iron	6.30
Calcium	5.00
Magnesium	2.90
Sodium	2.30
Potassium	1.50
Titanium	0.66
Carbon	0.18
Hydrogen	0.15
Manganese	0.11

Table 10.4 Element abundance in Earth's mantle. [Wikipedia 2021/Abundance_of_the_chemical_elements]

Element	%
Oxygen	45.00
Magnesium	23.00
Silicon	22.00
Iron	5.80
Calcium	2.30
Aluminum	2.20
Sodium	0.30
Potassium	0.30

data: Is it too far-fetched to assume different element creation processes for planets and stars? Is it far-fetched to assume different processes of element creation even on Earth itself?

10.2 THOUGHTS ON ELEMENT CREATION AND THE UNIVERSE

Having just talked about the different element abundances found in various environments, this point in the book is an opportunity to switch for a moment from the smallest to the biggest structures in our universe. We need to put in the groundwork for another "clean slate" which is unavoidable. We cannot avoid talking about electricity in the macroscopic world. Here, we are not talking about our daily life on Earth, instead we are talking about electricity in space. "Currently accepted theory" provides nothing on this topic. Although magnetic phenomena populate the surrounding space according to the Standard Model, electricity is nowhere to be seen. How all the magnetic fields around us are created is rarely discussed and the most obvious choice—electricity—is blissfully ignored. Instead, according to the Standard Model, gravity is the force dominating the Universe, together with explosions, collisions, shocks, and other actors on normal matter. If plasma is discussed, it is described as a "gas of ions"—not really accepting it as the fourth state of matter. Plasma is often considered to be the most abundant form of ordinary matter in the Universe. However, no consequences appear to be drawn from this fact. Instead, we have a model of creation of planetary systems, planets, and stars by accretion of matter through gravity which initiates a fusion process in objects' centers. If this is the case—independent of whether the accretion process would actually work—why are the element abundances on the surface of our Sun so radically different from the crust of our Earth, if everything was created from the same soup of material by accretion?

What consequences has this gravity-only mindset created instead for our view of the Universe? In the Standard Model the total mass–energy of the Universe contains 5% ordinary matter and energy, 27% dark matter, and 68% of an unknown form of energy known as dark energy. The presence of dark matter for example is implied in a variety of astrophysical observations that cannot be explained by accepted theories of gravity unless more matter is present. The rotation of galaxies is one example. Dark matter is called dark, because it does not appear to interact with observable electromagnetic radiation, and it is therefore undetectable by existing astronomical instruments. Dark energy, an unknown form of energy, was introduced to explain the measurements of an accelerated expansion of the Universe after the Big Bang.

This is the "reality" that the Standard Model has brought upon us: 95% of the Universe is something dark, something unknown, something unobservable and unmeasurable. If we consider the standards of science we presented in the introduction this so-called Standard Model has to go immediately. Neither does it follow scientific methodology nor the "trivium–quadrivium." It must be considered wrong, plain and simple. We need something new. This should immediately be followed by discarding the theory of the Big Bang, black holes, and other components of currently accepted theory that do not have a connection to anything observable or measurable. But wasn't the introduction of the Big Bang a result of the measurement of cosmic background radiation? Yes, that is the story, but the part of the story not often told is that predictions for cosmic microwave background radiation were far off, based on the theory of a Big Bang. Gamow predicted 20 K, Peebles went for 30 K:

> But the reporters had overlooked the fact that Penzias and Wilson had measured a temperature not of 30° K but 3.5° K [now 2.7 K]. This was considerably worse than it looked: the amount of energy in a radiation field is proportional to its temperature to the fourth power. The observed radiation had several thousand times less energy than Peebles or Gamow had predicted. Even by astronomers' standards, where factors of two are often chalked up to observational uncertainty, a disagreement of thousands of times bodes ill.
>
> Dicke told the New York Times that his group had predicted 10° K, which he considered acceptably close to the observations. (This figure is nowhere given in his published papers, so it's unclear where it came from.) And even 10° K yields a hundred-fold difference between the energy predicted and that observed. [Lerner 1991, 151–152]

Instead, it was a theological and philosophical decision to accept the Big Bang as the beginning of creation. As for black holes, they are already a mathematical impossibility,

concocted as parts of models that do not represent reality. Why black holes were introduced as a physical reality is beyond comprehension.

One other thing that some famous theories, created at the beginning and in the early
decades of the 20th century, have in common is the issue that they objectify concepts,
something to be avoided at all costs. It is simply not good science. This is another
reason to go back to before the introduction of those theories and start anew.

As a consequence, perhaps it is time to wipe the slate clean in astrophysics too.
The current model has led us into a dead end. This is not the place nor time to roll
out a new model for the Universe, but it should be noted that we see filament structures all around us. This fact points to the idea that electricity, charge separation,
double-layers, and other aspects of plasma play a much bigger role in the Cosmos than
previously considered. A double-layer is a structure in a plasma consisting of two
parallel layers of opposite electrical charge. The sheets of charge produce localized
excursions of electric potential, resulting in a relatively strong electric field. Consider
reading up on the so-called "Plasma Universe" [Plasma Universe 2021] or even the so-
called "Electric Universe" [Thornhill & Talbott 2007]—the latter widens the scope,
connecting mythology to natural electrical phenomena in the heavens.

10.3 PLASMA

We just talked about plasma. What is plasma?

'Plasma is a collection of charged particles that responds collectively to
electromagnetic forces' *(first paragraph in Physics of the Plasma Universe, Anthony
Peratt, Springer-Verlag, 1992).*

*A plasma region may also contain a proportion of neutral atoms and molecules,
as well as both charged and neutral impurities such as dust, grains and larger
bodies from small rocky bodies to large planets and, of course, stars.*

*The defining characteristic is the presence of the free charges, that is, the ions and
electrons and any charged dust particles. Their strong response to electromagnetic
fields causes behavior of the plasma which is very different to the behavior of
an unionized gas. Of course, all particles—charged and neutral—respond to a
gravity field, in proportion to its local intensity. As most of the Universe consists
of plasma, locations where gravitational force dominates that of electromagnetism
are relatively sparse.*

*Because of its unique properties, plasma is usually considered to be a phase of
matter distinct from solids, liquids, and gases. It is often called the 'fourth state of
matter' although, as its state is universally the most common, it could be thought of
as the 'first' state of matter.* [Johnson & Johnson 2011, chap. 3]

Plasma can be created when the outer electrons of an atom lose their connection to
the nucleus, the atoms being ionized. One way this can be achieved is by applying an
electric current that induces an electromagnetic force. It would seem likely that if we
have a high enough electromagnetic force available or if the force is active over a long

time, the quasi-inner electrons of a nucleus might be affected too as they reside closer to the nucleus while not being in the nucleus. As a consequence we would expect a strong electric discharge to be able to instantly induce nuclear decay—by moving or removing a quasi-inner electron. This might also be possible in a plasma, given more time. In Section 8.3 we mentioned that the weakening or the loss of the outer electron shield, a form of ionization, leaves the atom receptive to chemical as well as nuclear reactions. What should we take away from this section?

> Plasma appears to be an environment that facilitates nuclear reactions.

## CHAPTER 11

Simple nuclear reactions revisited

We discussed and used simple nuclear reactions in previous chapters of this book. It is now time to revisit those reactions. However, we should first take a look at radioactive decay.

11.1 WHAT IS THE CAUSE OF RADIOACTIVE DECAY?

This is what *Wikipedia* has to say about radioactive decay:

> *Radioactive decay is a stochastic (i.e. random) process at the level of single atoms. According to quantum theory, it is impossible to predict when a particular atom will decay, regardless of how long the atom has existed. However, for a significant number of identical atoms, the overall decay rate can be expressed as a decay constant or as half-life. The half-lives of radioactive atoms have a huge range; from nearly instantaneous to far longer than the age of the universe.* [Wikipedia 2021/ Radioactive_decay]

This clearly states that radioactive decay is an intrinsic property, a chance-based (random) decay at the atomic level. As we do not think that quantum theory applies to SAM we have to keep this as an open question. However, we have seen the importance of structure in the nucleus and how it influences its overall stability. The weak force from our point of view is not a real force since there is little more than an electromagnetic interaction between the nucleus and the environment. Furthermore, we cannot consider atoms to be energetically closed systems since they interact with their environment. They can exchange energies (e.g., through cosmic rays or photons). With some heavy radioactive elements, nuclear decay can be initiated through human interaction by providing PEPs, thus overcoming the structural integrity of the outer branches in the actinides. This represents an energetic hill that must be climbed before the natural downslope of energy release (exothermic) can occur. We will look at this in more detail in Chapter 13. In this chapter we will also describe other processes to initiate atomic decay with a predictable outcome.

Our statement here would be that radioactive decay is most likely not an intrinsic random process at the atomic level. For it to happen a set of circumstances has to be right. The reason we currently consider this process to be random lies with our inability to understand the process and to control the parameters because we do not know them yet. Based on SAM we think the key to understanding decay lies within the structure of the nucleus, amplified by the positioning of quasi-inner electrons once they appear. Once we are able to target a specific point on a nucleus and/ or apply energy to a specific atom, decay will no longer be considered intrinsically random.

11.2 β– DECAY/ELECTRON EMISSION

Looking back, the most prominent process was PEP capture in combination with β– decay. When overloading a previously stable nucleus with proton–electron pairs, we usually reach a threshold after adding two or three PEPs for the first elements. An exception was structurally required PEPs (Section 2.8). After adding two or three PEPs, the nucleus became unstable. It reacted by ejecting an inner electron from one PEP to relieve the nucleus of excess negative charge. In the process two proton–electron pairs combined to a deuteron on the nucleus and, of course, the superfluous electron was ejected. This also means that the proton–electron ratio changed in the nucleus as well as the number of deuterons. The count went up by one, which means we reached the next element. Looking at the energies and masses, as we did in the binding energy sections, the PEP emerges as a very curious thing. When an electron becomes part of a PEP, when it joins with a proton, the new pair gains mass, therefore becoming more resistant to forces. Energy needs to be expended (0.782287581 MeV) to join them— despite the attractive Coulomb force between them. The PEP is a somewhat "stressed" object, and when the nucleus itself accumulates negative charge through adding PEPs the charge is relieved by destroying a PEP—the weakest link. The destruction of a PEP releases energy. This energy as well as the change in binding energy going from one element to the next is the total amount of energy that is radiated away as heat or light or kinetic energy in β– decay. This released energy does not include the ejected electron itself. As seen before, β– decay is energetically possible if the loss in binding energy between father and daughter isotope does not exceed 0.782287581 MeV. If binding energy is gained through the decay step then β– decay is always possible.

11.2.1 Why is a PEP unstable outside a nucleus?

When a PEP is attached to a nucleus, the negative charge, the electron, acts at least partially as glue toward the nucleus. We can at least expect some concentration and localization of the electron toward a positive spot on the nucleus. In effect, we can say that the nucleus provides two anchor points. Nevertheless, it is still a stressed object. When a PEP is on its own, the electron associated with the proton has no localization, it is all around the proton—like a region (Fig. 11.1). Compared with an inner electron it is missing an anchor point (Section 2.14). This is not a preferred state for an inner electron since it detaches and either gets lost or converts to an outer one—the proton and electron becoming hydrogen-1 (Fig. 2.44).

We already pointed out that the PEP is a stressed object, that it releases energy when breaking into its components. The "free neutron" decay after roughly 15 minutes comes as no surprise. Only inside a nucleus, with a proper charge ratio and with two anchor points for the electron, can the PEP be stable.

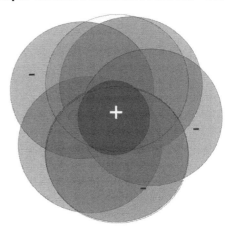

11.2.2 β– decay energy spectrum, anti-neutrinos, and spin

When discussing binding energy (Section 7.3), we established that the energy

Figure 11.1 PEP inner electron situation.

involved in a β– decay has two components: the change in binding energy and the released energy when the electron detaches from the proton. The sum of those two components is a discrete value. Now the interesting thing is that the β– decay kinetic energy is not a fixed value based on the sum of those two components, instead it is a continuous energy spectrum. The energy distribution typically goes up a bit and then goes down in something like part of a sine wave (Fig. 11.2). The expectation is to see a spike at the maximum energy level. However, that is not the case. The solution for this conundrum? An additional particle was invented, a so-called anti-neutrino, which rarely interacts with matter, has zero or very little mass, and little is known about, except what is required to fix the problem—in this case: carry the missing energy. This fateful invention by Wolfgang Pauli happened in 1930/1931. The particle was only later named the "anti-neutrino." All this did was to simply move the problem to something that cannot be measured easily. Problem solved! This reminds us of the supposed

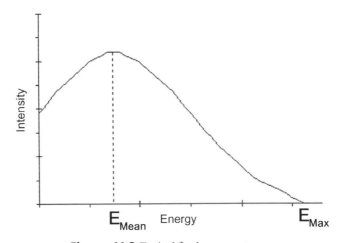

Figure 11.2 Typical β– decay spectrum.

purpose of dark matter and dark energy (Section 10.2). A simple, but effective trick that solves nothing in the end. Pauli himself called his step a "desperate remedy"—"... auf einen verzweifelten Ausweg verfallen ..." [https://www.symmetrymagazine.org/article/march-2007/neutrino-invention]

From our point of view, the beta decay is not a closed system (Section 11.1). This alone would create a continuous spectrum. Also, different electrons are being measured at different points in time of losing their energy. We also do not know if and how the electron changes from inner to outer or free electron. There is much more research to be done since we know next to nothing about this topic.

Another reason to invent a new particle (anti-neutrino) in this process is "spin" (Section 5.3). In the Standard Model the spin calculations of an electron, proton, and "neutron" don't add up unless we invent a new particle with the right "difference spin" and not much else. And this is exactly what Wolfgang Pauli did when he looked at conservation of energy, momentum, and angular momentum/spin of the β– decay. Now you know where the "anti-neutrino" and the "neutrino" came from in Section 1.5. A "neutron" is supposed to have a spin of ½. When it decays to a proton and an electron, the proton has a spin of ½ and the electron has a spin of ½. Simply adding the spins of a proton and an electron is incompatible with the spin of a "neutron." There is no path to get from A to B without inventing C. As we stated before, we do not think that spin is relevant for SAM as it is an artifact of a different model. Magnetic moment with direction and value however is relevant (Section 5.3). Again, we have reached an area requiring further research. We are also questioning the validity of those experiments in which "CP-symmetry violation" was declared.

11.3 DOUBLE β– DECAY/ELECTRON EMISSION

Double β– decay is a rare form of β– decay, where the energy levels of the nucleus prevent single β– decay—according to the Standard Model. The list of double β– decay-capable isotopes known so far includes calcium-48, germanium-76, selenium-82, zirconium-96, molybdenum-100, cadmium-116, tellurium-128, tellurium-130, xenon-136, neodymium-150, and uranium-238. Xenon-134 is a suspected candidate for double β– decay. The Standard Model gives little explanation beyond "the energy levels of the nucleus prevent a single β– decay."

What can the structure of the nucleus in SAM tell us about this issue? Calcium, germanium, selenium, tellurium, and neodymium, all have at least one missing element placed one deuteron count above the element. There also needs to be a sufficient number of protons (building material) to reach the next viable configuration. With selenium-82 for example we have two missing elements above it and bromine has no naturally occurring isotope with 82 protons. So, in this case we end up—including a correction for quasi-inner electrons—with a four-deuteron step upward while only two more outer electrons are counted. That actually is a quadruple β– decay/electron emission, where two ejected electrons are captured as quasi-inner electrons and not viewed as decay steps. Again, the structure of the nucleus in SAM delivers interesting explanations.

If we look at calcium-46 (Fig. 11.3) and calcium-48 (Fig. 11.4) we expect them to

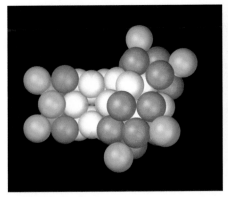

Figure 11.3 Calcium-46, which has a long half-life.

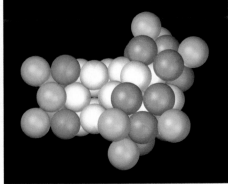

Figure 11.4 Calcium-48, which has a long half-life.

decay quickly because the additional PEPs push the ratio of negative to positive charge in the nucleus beyond the typical threshold for β– decay. However, the isotopes are considered observably stable or as having a *very long* half-life. Instead, when they finally decay, the result is titanium-46 (assumed) and titanium-48 (measured), respectively. Something—at least one missing unstable element—prevents decay until the conditions are right for a bigger step.

11.4 β+ DECAY/ELECTRON CAPTURE

Another reaction we used was β+ decay or electron capture. An (outer) electron is captured by a proton in the nucleus being a single proton or part of a deuteron, converting both into a PEP. This changes the deuteron/single proton count; the result is an element stepdown maybe combined with a resettling step. We have reached the previous element. In Section 7.3 we looked at the example of sodium-22 becoming neon-22. We established that in this case the nucleus first rearranged, which released 3.625330276 MeV. Some of this energy (0.782287581 MeV) was used to supply the energy to put a proton and an electron together to form a PEP. The remaining energy (2.843042695 MeV) is radiated (e.g., as kinetic energy of the resettling or as γ rays). What does this look like? See Table 11.1.

The big γ decay of 1.27457 MeV is probably the result of relocation of the PEP on the nucleus. The amount of energy might be proportional to the distance it has to move. Very interesting are the 2×0.511-MeV γ steps. The value of 0.511 MeV also represents the energetic equivalent of the rest mass of an electron. However, no extra electrons are involved in this calculation as the electron was either taken from the outer electrons or one outer electron had to be ejected. The number of electrons is therefore balanced in any case. As the "summary" column in Table 11.1 shows, we always reach our initial energy amount.

Here is another case, this time we are looking to radiate 1.504 MeV (Table 11.2).

The remaining 89.25% are for a β– decay to calcium-40. The 1.460822-MeV γ emission here has a higher value than the 1.27457 MeV for sodium-22. This may be

Table 11.1 Na-22 to Ne-22 β+ energies.

Probability × 100	Mode	Energy	Probability × 100	Mode	Energy	Summary energy
0.00098%	EC trans.	2.843 MeV				2.843 MeV
0.05500%	Gamma	2 × 0.511 MeV				2.843 MeV
	β+ trans.	1.82102 MeV				
99.94000%	Gamma	1.27457 MeV	90.30%	Gamma	2 × 0.511 MeV	2.843 MeV
				β+ trans.	0.54644 MeV	
			9.64%	EC trans.	1.56844 MeV	2.843 MeV

EC trans., electron capture transition.

Table 11.2 K-40 to Ar-40 decay energies.

Probability × 100	Mode	Energy	Summary energy
0.00100%	β+ trans.	0.4829 MeV	1.504 MeV
	Gamma	2 × 0.511 MeV	
0.20%	EC trans.	1.504 MeV	1.504 MeV
10.55%	EC trans.	0.044 MeV	1.504 MeV
	Gamma trans.	1.460822 MeV	

EC trans., electron capture transition.

explained by relocation of the PEP which has to travel a greater distance and can gain more kinetic energy.

Finally, we will look at one example with a very-low-energy β+ decay/e-capture (Ca-41 to K-41), expecting to see 0.42163 MeV (Table 11.3).

The so-called electron capture event might just be an indicator for the in-place conversion of a proton into a PEP, with minimal resettling. We would argue that a β+ decay/e-capture is energetically possible if the gain in binding energy between parent and daughter isotope is at least 0.782287581 MeV. Below that, there is not enough energy available to attach the electron to the proton, creating a PEP.

Table 11.3 Ca-41 to K-41 decay energies.

Probability × 100	Mode	Energy	Summary energy
100.00%	EC trans.	0.42163 MeV	0.42163 MeV

EC trans., electron capture transition.

11.4.1 Positrons and neutrinos

As discussed in the introduction to simple nuclear reactions (Section 1.5), instead of an electron being captured by the nucleus during β+ decay, "currently accepted theory" sees a positron being ejected, as well as a neutrino. They also would make us believe that the 2 × 0.511 MeV of gamma rays sent out at roughly 180° are the result of an ejected positron annihilating with an electron (Where does the positron reside in the nucleus? Or, if the positron does not reside in the nucleus, why does it appear at the right moment in the right place?), resulting in the 0.511 MeV − −0.511 MeV = 1.22 MeV energy we observe. Instead, we are of the opinion that the positron is not a particle leaving, but what we actually see is one electron entering the nucleus. The view of this incoming electron as an outgoing positron is a view back in time, like a film running backward. On the basis of logic we have, at least in this case, to reject the notion of a positron actually being a real, ejected "object." The Standard Model can't allow an electron to enter the nucleus, therefore it needs to be destroyed before the electron can enter it. However, there is the previously mentioned observation of an energy release of 2 × 0.511 MeV of gamma rays sent out at roughly 180°. We think this is the signature of the destruction of a deuteron. The value of 0.511 MeV (the rest mass of an electron) could be coincidental, although it would be a strange coincidence. Interestingly, the Standard Model sees an electron capture event instead of positron ejection if the energy levels are low enough. Where does the electron go? It forms a neutron with a proton in the nucleus, just like it is the case in SAM. In SAM it is always an electron entering the nucleus. The only difference is: We are either converting a single proton to a PEP in place or destroy a deuteron by creating two PEPs close to each other, and then usually one of the PEPs moves.

The conversion of energy, momentum, and angular momentum (spin), forces—according to the Standard Model—the existence of the neutrino to fix the spin calculation. Again, we think this is irrelevant because discrete spin numbers are an artifact of a different model. The decay spectrum of the β+ decay is continuous in the same way as β− decay. The values are slightly shifted because of the Coulomb force between the nucleus and the electron, either incoming (β+) or outgoing (β−).

11.5 DOUBLE β+ DECAY/ELECTRON CAPTURE

This is a rare form of β+ decay/electron capture, where the structure or energy levels prevent a single β+ decay—according to the Standard Model. The list of double β+

decay/electron capture capable isotopes known so far include krypton-78, xenon-124, and barium-130. With krypton and xenon it is a symmetry issue, it can only complete double β+. It then ends up at a missing, unstable element, decays further, and has to correct for the proper quasi-inner electron count. The stepdown in deuterons is actually four, but the stepdown in outer electrons is only two as the other two are sourced from quasi-inner electrons. With barium-130 the single β+ would create cesium-130, which is too light and not naturally occurring. It can only do a double step to reach xenon-130. The requirement to correct for a different number of quasi-inner electrons doesn't arise in this case.

11.6 PEP EMISSION

We also have seen PEP emission when overloading nuclei with additional PEPs. At some point the connection to the nucleus is so weak that they "float" away very easily—creating halo PEPs (see Section 4.1) or decaying into a proton and electron. This relates to the so-called "beta-delayed proton decay." One possible example is a PEP emission when going from beryllium-11 to beryllium-10. It is suggested by Standard Model proponents that instead we see first a β− decay to boron-11 and then a proton emission to beryllium-10. But boron-11 is stable. Why would it further decay? This makes total sense in SAM since the halo PEP in beryllium-11 is already close to breaking apart into a proton and an electron as it is not being stabilized as part of the nucleus. We simply see the electron emission and the proton emission of what once was a halo PEP. Most likely, beryllium-11 never becomes boron-11 as an intermediate step since the proton is not really part of the nucleus structure.

11.7 PROTON EMISSION

When discussing sulfur-32/phosphorus-32 (Section 2.11), we briefly mentioned another process that also resulted in an element stepdown. A moderately fast PEP is captured by a nucleus, but instead of attaching to the nucleus as happens in the PEP capture process, the proton of the pair (or another proton?) is ejected again. The electron of the PEP is integrated into the nucleus which makes the result identical to a β+ decay, a stepdown to the previous element. The PEP could be considered to be just a carrier for the electron. In a sense this could be considered to be a proton emission. With PEPs of a lower energy level, we reach sulfur-33. This is how the Standard Model describes the process. From the viewpoint of SAM a different story emerges. If we compare sulfur-32 and phosphorus-32 by the numbers (Appendix H), the only difference we see is that the outer electron count goes down while the quasi-inner electron count goes up. All the moderately fast PEP does is to kinetically trigger a resettling of the nucleus which then gives room for a quasi-inner electron. This results in an outer electron being pulled between the endings. The PEP continues on its course, maybe having decayed into a proton and an electron itself. There is no proton emission in this case since this is a purely kinetic, destructive process involving a PEP.

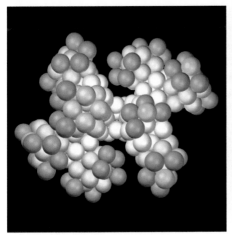

Figure 11.5 Polonium-210.

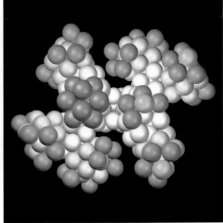

Figure 11.6 Lead-206.

11.8 α DECAY

We encountered the α decay at several instances. We think it happens when two deuterons are released from the nucleus at the same time in close proximity (e.g., a four-ending). They then recombine into the tetrahedral structure of the helium-4 nucleus, a very much energetically preferred configuration.

What would be the precondition for α decay? There needs to be enough structure in the nucleus for four-endings, two-endings, or some PEPs on different branches to be close together. This is definitely the case for heavy nuclei. When those endings get released they are close enough to easily combine. However, the option for this type of decay already exists once the middle branches start to grow, basically around barium, where the middle branches are becoming recognizable. Not surprisingly, α decay starts to happen in PEP-deficient nuclei just before the PTE section we described above—with tellurium.

There are heavier elements like lead where one of the middle branches is just a carbon nuclet without cappings. Therefore, we do not see much α decay in elements like lead, at least with naturally occurring isotopes.

Probably the easiest way to see α decay in action is with the step from polonium-210 (Fig. 11.5) to lead-206 (Fig. 11.6). The middle carbon nuclet loses a four-ending which recombines into an α particle.

11.9 α CAPTURE

We asked before if there is such an event as α capture. The answer is yes it exists. However, the tetrahedral structure is such a preferred configuration. Why would that be given up easily? The only option to achieve this would be to shoot a helium-4 nucleus with such high energy onto a target nucleus that it gets destroyed (by impact!) and components are picked up by the target nucleus.

Lord Rutherford and his group of scientists were the first to produce and detect artificial nuclear transmutations in 1919. They bombarded nitrogen in the air with α particles emitted during the decay of polonium-214. The transmutation reaction involved the absorption of an α particle by the nitrogen-14 nuclei to produce oxygen-17 and a proton (a hydrogen nucleus).

Remember in the beginning (Sections 1.4 and 2.2) we had the issue with elements with five or eight protons. They are not possible. As a result, combined with the issues around potassium, we had to conclude that elements are created by putting together bigger chunks (at least of deuteron size, possibly up to oxygen), as is proven by this very first transmutation experiment. Just adding a PEP, one at a time, is sometimes not enough. An option to achieve this is a *capture*.

11.10 PEP CAPTURE

Since PEP capture plays an important role we will discuss it in more detail. It is apparent that what happens when a PEP hits a nucleus depends, first, on where it hits, second, whether the nucleus is particularly receptive to PEP capture, and third, with what (kinetic) energy it hits. The energy classification ranges with some detail from cold (0.0–0.025 eV) to thermal (0.025 eV) up to ultrafast (>20 MeV). The most used classification is simpler:

- slow group (0.025 eV–1 keV); and
- fast group (1 keV–10 MeV).

A slow (thermal) proton–electron pair ("neutron") shows behavior that implies the PEP, when delicately offered to the nucleus, will find a "vacant spot." When offering a fast proton–electron pair to the nucleus we see that other spots may be hit and the PEP may react by being absorbed as an extra "neutron" or trigger a β– decay process by emitting one electron while combining with another proton–electron pair in the nucleus. Offering a fast PEP may also result in hitting the "backbone" which is by its nature unable to accept a PEP due to being completely "neutralized." This will result in the proton–electron pair showing scattering effects. Offering a proton–electron pair (PEP) with too much energy may result in the destruction of the nucleus which would be the case when a high-energy cosmic ray hits a nucleus. Particle accelerators are also able to do this.

Let's have a look at zirconium-88. It is a synthetic isotope which looks in SAM like Figure 11.7, with two four-endings.

Those two four-endings are not enough to balance the lithium nuclets. It will decay with a half-life of 83.4 days. If we offer thermal PEPs to zirconium-88, they will find the four-endings and create five-endings. This then is a stable configuration (Fig. 11.8).

The faster the PEPs, the more the type of reaction relies on where they hit the nucleus. If the spots on the four-endings are taken by PEPs (now five-endings), the PEPs will next attach to the lithium nuclets (Fig. 11.9).

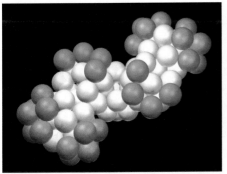

Figure 11.7 Zirconium-88 (*view from top*).

Figure 11.8 Zirconium-90 (*view from top*).

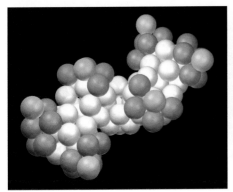

Figure 11.9 Zirconium-94 (*view from top*).

Figure 11.10 Zirconium-96 (*view from top*).

The current assumption is that the lithium nuclets can take only one extra PEP before making the nucleus unstable. Therefore, zirconium-94 has to look like Figure 11.9. More PEPs would most likely attach to the five-endings, a situation that is also not stable, but zirconium-96 has a high half-life of 2.0×10^{19} years (Fig. 11.10). Here we basically follow the rules established in Section 4.1.

If, on the other hand, we hit the back-side of zirconium with a PEP, what is there to attach to (Fig. 11.11)?

Even for the back-side it might be possible to attach to the lithium nuclets. However, the bulk of the back-side of the nucleus is "neutralized" and a PEP either could not attach or could only attach very weakly. The probability of PEP capture is also called "capture cross-section," measured in "barns." It was found that zirconium-88 has a very high capture cross-section while other zirconium

Figure 11.11 Zirconium-90 (*backside view*).

isotopes have a low capture cross-section. Looking at the structure it becomes clear why that might be the case since some attachment points are already occupied.

11.11 PROTON CAPTURE

There is one more process that we have not talked about and that is rarely discussed: proton capture. Indirectly, we have seen proton capture already. When a PEP is captured by a nucleus, followed by electron emission (β– decay) because the nucleus became unstable, the result is a proton added to the nucleus. However, we wonder, is there a way for a nucleus to directly capture a proton? Again, one direct example we have already seen, although a very unstable one, is when lithium-7 is hit with protons and absorbs one. However, eight spheres are not a stable configuration so lithium-7 plus the proton decays into two helium-4 nuclei (Section 2.1). So, the question is whether it is possible to have proton capture with a stable end result?

As with proton–electron pair (PEP) capture we must consider some aspects:

- the general receptiveness of a nucleus for protons;
- the location the proton hits;
- the angle the nucleus is hit from; and
- the energy of the proton.

In general, an overall positive nucleus should not be very receptive to a proton because of Coulomb's law—causing the so-called Coulomb barrier. However, we have already seen that there are spots on the nucleus that are less positive than others. Also, if an atom is bound to other atoms (e.g., in a metal lattice), the positive connection points are covered by electrons from other atoms. So, there might be holes in the Coulomb barrier around the nucleus under some circumstances. Hitting close to such a point might be helpful for proton capture. Getting through to such a point also depends on the energy of the proton. Just enough energy to get through the barrier might be helpful too. With energies in excess of 50 MeV the proton might even act like a bullet, piercing through the nucleus and destroying it in the process.

Which structures of a nucleus could be considered receptive to proton capture? One example is boron-11 (Fig. 11.12) or in more general form the boron ending (e.g., in aluminum-27) (Fig. 11.13).

With just a proton, boron-11 will become ionized carbon-12 (Fig. 11.14) and aluminum-27 will become ionized silicon-28 (Fig. 11.15), both completing the icosahedron structure. However, there is a difference. Whereas with boron-11 a new deuteron needs to be created, this is not the case with the boron ending. There the proton is just added to the existing five deuterons to create a carbon nuclet. It is much more likely that boron-11 will react to a proton with an α decay (analog to rare boron-12 α decay mentioned in Section 2.1) than the boron ending. However, this is another research topic.

The second, more likely, option for proton capture has to do with the large percentage of elements and isotopes that carry extra PEPs for structural reasons. These PEPs are able to connect to a proton if it comes close enough. The five-endings are a prime

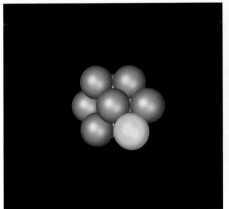

Figure 11.12 Boron-11.

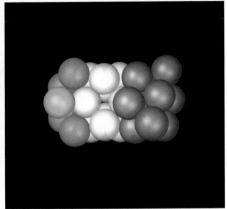

Figure 11.13 Aluminum-27.

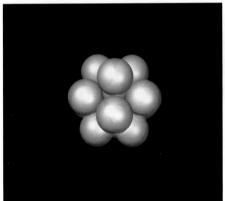

Figure 11.14 Carbon-12.

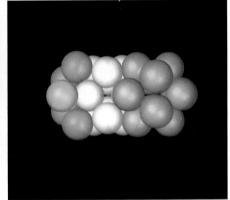

Figure 11.15 Silicon-28.

candidate for this. They create a deuteron which changes the element to the next one in an ionized form. Often that means that the five-ending is converted into a lithium nuclet. An illustrative example would be sodium-23 capturing a proton and directly converting into magnesium-24 (Figs. 11.16, 11.19, and 11.20). The reaction is similar to a β– decay. However, there is no change in the inner electron number because an inner electron is already available due to the presence of the PEP. All that is lacking in the nucleus to make it to the next element is a proton. The missing outer electron must be stolen from somewhere else.

The normal path (A) from sodium-23 to magnesium-24 would work through PEP capture on the five-ending followed by resettling of the PEP including a β– decay (Figs. 11.16, 11.17, 11.19, and 11.20). The electron that is ejected from the nucleus during β– decay could become the missing outer electron. The PEP could also (B) have been captured by the lithium nuclet, then it would have to transition to the five-ending first (Figs. 11.16, 11.18–11.20).

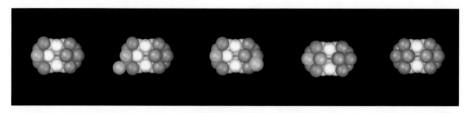

Figure 11.16
Sodium-23.

Figure 11.17
(A) PEP capture
on five-ending,
followed by
resettling, and
β– decay.

Figure 11.18
(B) PEP capture
on lithium
nuclet, isomeric
transition, and
β– decay.

Figure 11.19
Proton capture or
status after (A/B)
and β– decay.

Figure 11.20
Magnesium-24.

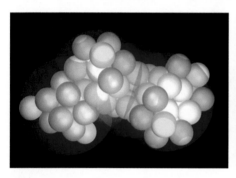

Figure 11.21 Nickel-62 with nuclet-
bounding spheres.

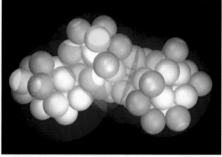

Figure 11.22 Copper-63 with nuclet-
bounding spheres.

In many other cases this reaction of a five-ending with a PEP would create a lithium nuclet in the wrong place on the nucleus. This leads to spallation and the whole nucleus resettles or disintegrates.

If we apply this proton capture logic to nickel, then it is conceivable that exposing, say, nickel-62 (Fig. 11.21) to ionized hydrogen-1 (protons) with sufficient energy would cause a nuclear reaction (p-capture) to copper-63 (Fig. 11.22). A nickel-58 with two four-endings would be less receptive to proton capture (Fig. 11.23).

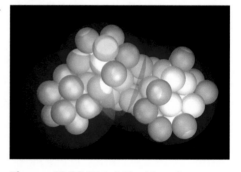

Figure 11.23 Nickel-58 with nuclet-
bounding spheres.

If the result is not a stable configuration, the nucleus will respond. Unfortunately, an unstable outcome is more likely than a stable one.

This resettling (self-organizing) of the nucleus can then result in a stable nucleus with no further emissions. Resettling can also cause a β– or β+ decay step (change in inner/outer electrons). Under very unstable conditions the nucleus can react by emitting bigger parts of the nucleus. This will be discussed further in Chapter 13.

The topic of proton capture is our most important lead to explaining a class of reactions that are often recognized as something like "this must have happened," but the Standard Model does not allow for it. This of course needs to be studied in much more detail. How and why a proton is captured in all its details is still somewhat unclear. Based on SAM we can make some educated guesses such as the nucleus needs to be receptive by having available PEPs or boron endings in its structure. Also, the environment must be right, which means positive spots on the nucleus must be neutralized as much as possible.

In the Standard Model, proton capture is only used to describe how certain elements are created during cosmological events. Isotopes that are too light (having too few PEPs) are sometimes found in nature (e.g., helium-3 or beryllium-7, believed to be created just after the Big Bang) and pose questions about how these isotopes are still observed, since they are unstable as a rule. The current view in the Standard Model is that a rapid absorption of many neutrons can lead to the creation of heavier elements in places/events such as neutron star mergers or supernovae. From our point of view, the question is whether neutron stars actually exist—let alone merge—and whether we understand correctly what supernovae signify in the life cycle of a star (Section 10.2).

Instead, we want to ask whether proton capture can happen on Earth—in nature and the laboratory. We will see where the evidence leads us . . .

11.12 ELEMENT BUILDUP REVISITED

If proton capture is possible and if proton capture happens on Earth, then there is another option to originate a new element. Instead of adding a PEP and another PEP and then a β– decay step, it should be possible to add a PEP and then a proton or first a proton and then a PEP. The order depends on the isotope being initially receptive to protons. The missing outer electron would be stolen from somewhere else.

To summarize: moving from one element to the next and considering the deuteron count as a base, we can follow several paths to the next element:

A Add a PEP, and another PEP, then lose an inner electron (normal β– decay, C12 → C13 → C14 → N14).
B Add a PEP, then a proton, then add an outer electron (B10 → B11 → C12).
C Add a proton, then add a PEP, then add an outer electron (in rare cases such as Hydrogen-2 → Helium-3 → Helium-4).
D Add a deuteron, then add an outer electron.

In some special cases, options B and C will not work because an unstable element is created in the process. One example would be starting with calcium-40, where we already have a PEP added (it is required), then we add a proton. The proton pulls in an electron as an outer electron, then the outer electron is further pulled in as a quasi-inner electron. The lithium structure we just created while adding the proton

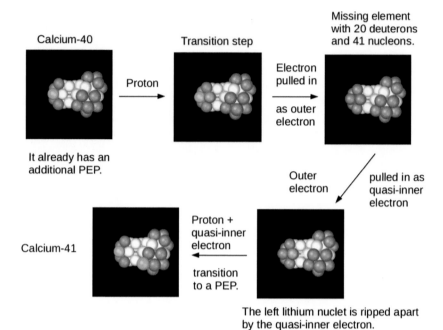

Figure 11.24 Attempting to move from calcium-40 to the next element by adding a proton.

(a step to the next element) cannot hold the quasi-inner electron in place. It falls back to a five-ending, and we end up with a second PEP on one of the lithium nuclets (i.e., calcium-41) (Fig. 11.24).

In order to bypass those unstable elements, we need to apply bigger steps. Double β– decay is apparently one of nature's solutions.

11.13 β DECAY AND NUCLEAR STRUCTURE

As stated before (Section 11.1), we think that nucleus structure determines which decay channels exist. It has everything to do with the structure of the nuclei involved. Let's look at two examples:

1 Fluorine-18 β+ decay to oxygen-18. According to the literature there is one direct path via e-capture and one β+ decay with an isomeric transition. During the e-capture step, the proton on the right side directly captures an electron and transforms into a PEP (Fig. 11.25, *top*) without movement (Fluorine-18B). The β+ decay path with the isomeric transition first breaks up the single deuteron on the right side (releasing 1.022 MeV in the process) and then moves the newly created PEP to the new position on the left (Fluorine-18A). In Section 9.7 we came to the conclusion that two versions of fluorine-18 can exist and fluorine-18A might be

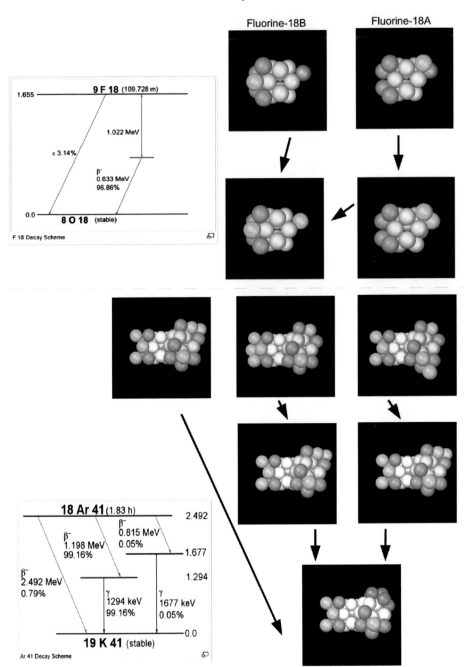

Figure 11.25 Comparison of decay schemes with nuclear structure (fluorine/argon).

a preferred state because of the smaller minimal bounding sphere. Figure 11.25 (*top*) compares the decay schemes with the structural changes.

2 Argon-41 β– decay to potassium-41. According to the literature there exists one direct path and two decay paths with isomeric transitions. In the direct path turnout (only β– decay), a PEP releases its electron in place and directly converts the five-ending into a lithium nuclet by creating a deuteron. As for the isomeric transition variants, there are two options since there are two other five-endings on the base carbon nuclets that can hold a PEP. This PEP moves first to the right position as a PEP and then loses the electron. The energetic differences in the γ ray are most likely the result of the different distances the PEP has to travel to reach the target five-ending. Figure 11.25 (*bottom*) again compares the decay schemes with the structure.

What should we take away from these two examples? Fluorine has one ending where the change happens, but two configurations. Argon has three five-endings that could turn into a lithium nuclet, we have therefore three starting positions where the additional PEP could be. The two isomeric transitions represent the relocation of the PEP to the right five-ending. We think the decay schemes and the structural options, once fully researched, will correlate strongly.

CHAPTER 12

Fusion

12.1 VIABILITY OF FUSION

We already talked about fusion when we discussed simple nuclear reactions. Every capture in a sense is some kind of fusion. However, the typical definition of fusion reads:

> *Nuclear fusion is a reaction in which two or more atomic nuclei are combined to form one or more different atomic nuclei and subatomic particles (neutrons or protons).* [Wikipedia 2021/Nuclear_fusion]

This excludes e-capture and PEP capture since neither are atomic nuclei; proton capture barely qualifies but is excluded.

The biggest absorption/capture we have considered so far is α particle capture, and we think it is not very likely in its pure, slow form for structural and energetic reasons. However, you can shoot α particles at something with high energy and see if at least some fragments stick to the target. In 1917 Ernest Rutherford and his team were able to accomplish the transmutation of nitrogen into oxygen using α particles directed at nitrogen-14, which resulted in oxygen-17 and an emitted proton. The result of a real (slow) α particle capture would have been β+ decay and oxygen-18 in this case. However, that was not observed.

In between the proton and helium-4 we have deuterium. Deuterium fusion in rare cases can be very easy, like snapping on a two-ending in the capping phase of a carbon. However, most likely a violent rearrangement of the nucleus follows such a fusion step, perhaps even fission as a secondary step.

One question remains: Is fusion possible with even bigger nuclei? Due to the nature of the backbone structure, we can say that there is a limit to how many protons can be added to a nucleus for the fusion reaction to take place in such a way that the result is a larger stable element. What we mean by this is that there would be a maximum number of protons and inner electrons we can add to an existing nucleus for the reaction to be "correct" and the result to be stable. This is true especially for very heavy nuclei.

For example, two carbon nuclei appear to be able to resettle into a magnesium nucleus. However, adding an iron nucleus to a silver nucleus would not be a conceivably viable reaction, and even if this reaction were to take place, which it most likely would not due to energetic and structural reasons, it would not work out and at the very least it would yield fragments in a chaotic manner.

Fusing a small nucleus with another small, or a moderate size, nucleus seems therefore to be a good theoretical option, but what does nature tell us about this type of fusion?

When we looked at the element and isotope abundance values for Earth (Section 10.1) we noticed that about half of Earth's crust is made up of the element oxygen. We predominantly see oxides or derivatives of that in the geological record. Could oxygen play an important role in fusion processes and can we see this reflected in the geological record and the abundance of the elements? Do these reactions even make sense when we look at the structures?

Here are a few reactions we might consider. When we look at them for some time we see a trend in abundances. The most available elements plus an oxygen is the next "most abundant element." A complete cycle-of-eight (16 protons) is added to an element turning it into the equivalent element in the next cycle-of-eight. Some examples:

Carbon-12   + Oxygen-16 → Silicon-28
Oxygen-16   + Oxygen-16 → Sulfur-32
Fluorine-19 + Oxygen-16 → Chlorine-35
Sodium-23   + Oxygen-16 → Potassium-39
Magnesium-24 + Oxygen-16 → Calcium-40

All these steps reflect the similarity of the product before and after as well as the abundances that more or less keep showing the same relation.

In the 20th century the biologist C. Louis Kervran (1901–1983) studied nuclear reactions in biology and wrote books and papers on the topic. Next to proton capture and α particle absorption he also points out many suspected reactions with oxygen-16. The general idea here is that elements, regardless of their size (number of protons), can react with oxygen-16 and fuse together, creating an element that is by default 16 protons larger, usually translating into a stepup by eight elements [Kervran 1980].

Oxygen is a good candidate, but other elements might be possible for fusion too, depending on the environment and circumstances. Interestingly, a metal–oxygen fusion reactor has already been patented [Swartz 2017].

12.2 ENERGY CALCULATIONS FOR FUSION

Now let's have look at some of the fusion reactions described earlier as well as others suggested from the viewpoint of the binding energy involved (Tables 12.1–12.19).

Table 12.1 Binding energy for deuterium/deuterium fusion.

Fusion reaction components	Total BE (MeV)	SAM BE (MeV)	Difference BE (MeV)
Hydrogen-2	2.23	2.23	0
Hydrogen-2	2.23	2.23	0

Fusion reaction components	Total BE (MeV)	SAM BE (MeV)	Difference BE (MeV)
Hydrogen-2 + Hydrogen-2	4.45	4.45	0
Helium-4	28.3	26.7	1.6
Helium-4 – (Hydrogen-2 + Hydrogen-2)	23.85	22.25	1.6

Deuterium–deuterium fusion is the holy grail of fusion experiments and looking at the difference in binding energy it is easy to understand why: 23.85 MeV can be gained per atom by this fusion step (Table 12.1).

Table 12.2 Binding energy for lithium/helium fusion.

Fusion reaction components	Total BE (MeV)	SAM BE (MeV)	Difference BE (MeV)
Lithium-7	39.25	42.28	–3.03
Helium-4	28.3	26.7	1.6
Lithium-7 + Helium-4	67.55	68.98	–1.43
Boron-11	76.21	77.88	–1.67
Boron-11 – (Lithium-7 + Helium-4)	8.66	8.9	–0.24

The step from lithium and helium to boron is again an energy gain (Table 12.2).

Table 12.3 Binding energy for carbon/helium fusion.

Fusion reaction components	Total BE (MeV)	SAM BE (MeV)	Difference BE (MeV)
Carbon-12	92.16	91.23	0.93
Helium-4	28.3	26.7	1.6
Carbon-12 + Helium-4	120.46	117.93	2.53
Oxygen-16	127.62	126.83	0.79
Oxygen-16 – (Carbon-12 + Helium-4)	7.16	8.9	–1.74

The binding energy of carbon-12 and helium-4 combined is less than the binding energy of oxygen-16. As a result, 7.16 MeV can be gained by this fusion process per atom (Table 12.3).

Table 12.4 Binding energy for carbon/oxygen fusion.

Fusion reaction components	Total BE (MeV)	SAM BE (MeV)	Difference BE (MeV)
Oxygen-16	127.62	126.83	0.79
Carbon-12	92.16	91.23	0.93
Oxygen-16 + Carbon-12	219.78	218.06	1.72
Silicon-28	236.54	238.08	−1.54
Silicon-28 − (Oxygen-16 + Carbon-12)	16.76	20.02	−3.26

The binding energy of carbon-12 and oxygen-16 combined is less than the binding energy of silicon-28. As a result, 16.76 MeV can be gained by this fusion process per atom. Also, one deuteron is split into a single proton in the additional carbon nuclet and a PEP to form the five-ending (Table 12.4).

Table 12.5 Binding energy for oxygen/oxygen fusion.

Fusion reaction components	Total BE (MeV)	SAM BE (MeV)	Difference BE (MeV)
Oxygen-16	127.62	126.83	0.79
Oxygen-16	127.62	126.83	0.79
Oxygen-16 + Oxygen-16	255.24	253.66	1.58
Sulfur-32	271.78	273.68	−1.9
Sulfur-32 − (Oxygen-16 + Oxygen-16)	16.54	20.02	−3.48

The binding energy of oxygen-16 and oxygen-16 combined is less than the binding energy of sulfur-32. As a result, 16.54 MeV can be gained by this fusion process per atom (Table 12.5).

Table 12.6 Binding energy for fluorine/oxygen fusion.

Fusion reaction components	Total BE (MeV)	SAM BE (MeV)	Difference BE (MeV)
Fluorine-19	147.8	155.75	−7.95
Oxygen-16	127.62	126.83	0.79
Fluorine-19 + Oxygen-16	275.42	282.58	−7.16
Chlorine-35	298.21	302.6	−4.39
Chlorine-35 − (Fluorine-19 + Oxygen-16)	22.79	20.02	2.77

The binding energy of fluorine-19 and oxygen-16 combined is less than the binding energy of chlorine-35. As a result, 22.79 MeV can be gained by this fusion process per atom. Also, one deuteron is split into a single proton in the additional carbon nuclet and a PEP (Table 12.6).

Table 12.7 Binding energy for sodium/oxygen fusion.

Fusion reaction components	Total BE (MeV)	SAM BE (MeV)	Difference BE (MeV)
Sodium-23	186.56	191.35	−4.79
Oxygen-16	127.62	126.83	0.79
Sodium-23 + Oxygen-16	314.18	318.18	−4.00
Potassium-39	333.72	338.2	−4.48
Potassium-39 − (Sodium-23 + Oxygen-16)	19.54	20.02	−0.48

The binding energy of sodium-23 and oxygen-16 combined is less than the binding energy of potassium-39. As a result, 19.54 MeV can be gained by this fusion process per atom. Also, one deuteron is split into a single proton in the additional carbon nuclet and a PEP (Table 12.7).

Table 12.8 Binding energy for magnesium/oxygen fusion.

Fusion reaction components	Total BE (MeV)	SAM BE (MeV)	Difference BE (MeV)
Magnesium-24	198.26	198.03	0.23
Oxygen-16	127.62	126.83	0.79
Magnesium-24 + Oxygen-16	326.22	324.86	1.02
Calcium-40	342.05	344.88	−2.93
Calcium-40 − (Magnesium-24 + Oxygen-16)	15.83	20.02	3.95

The binding energy of magnesium-24 and oxygen-16 combined is less than the binding energy of calcium-40. As a result, 15.83 MeV can be gained by this fusion process per atom. Also, one deuteron is split into a single proton in the additional carbon nuclet and a PEP to form the five-ending (Table 12.8).

Table 12.9 Binding energy for silicon/oxygen fusion.

Fusion reaction components	Total BE (MeV)	SAM BE (MeV)	Difference BE (MeV)
Silicon-28	236.54	238.08	−1.54
Oxygen-16	127.62	126.83	0.79

Fusion reaction components	Total BE (MeV)	SAM BE (MeV)	Difference BE (MeV)
Silicon-28 + Oxygen-16	364.16	364.91	−0.75
Missing element 21/43 + PEP	N/A	376.03	N/A
Missing element 21/43 + PEP − (Silicon-28 + Oxygen-16)	N/A	11.12	N/A
Calcium-44	380.96	371.58	9.38
Calcium-44 − (Silicon-28 + Oxygen-16)	16.8	6.67	10.13

Silicon has 13 deuterons and a single proton. Oxygen has 8 deuterons. The fusion of both would initially yield 21 deuterons. The (missing, unknown) base element with 21 deuterons has only 43 protons, so we would expect an additional PEP on this configuration to reach the required 44 protons. The indicated missing element is unstable, even in its base configuration. It decays with β+ steps to calcium-44. The binding energy of silicon-28 and oxygen-16 combined is less than the binding energy of calcium-44. As a result, 16.8 MeV can be gained by this fusion process per atom, but not in a single step. Also, it is noticeable that the overall SAM binding energy gain through fusing with oxygen is less (11.12 MeV) than we have seen before (20.02 MeV). The reason is that the base carbon of the oxygen in this case gets distributed and is not kept intact as in the other cases (Table 12.9).

Table 12.10 Binding energy for calcium/carbon fusion.

Fusion reaction components	Total BE (MeV)	SAM BE (MeV)	Difference BE (MeV)
Calcium-44	380.96	371.58	9.38
Carbon-12	92.16	91.23	0.93
Calcium-44 + Carbon-12	473.12	462.81	10.31
Chromium-56	488.5	476.35	12.15
Chromium-56 − (Calcium-44 + Carbon-12)	15.38	13.54	1.84
Iron-56	492.26	491.73	0.53
Iron-56 − (Calcium-44 + Carbon-12)	19.14	28.92	−9.78

The reaction for calcium/carbon fusion is complicated. The initial result actually—according to deuteron and single proton count—is chromium-56. However, that is unstable and β− decays with a half-life of 5.94 minutes to manganese-56. And manganese-56 with a half-life of 2.5789 hours finally decays to iron-56 by another β− step. The binding energy of calcium-44 and carbon-12 combined is less than the binding energy of iron-56. As a result, 19.14 MeV can be gained by this fusion process per atom, although not in a single step (Table 12.10).

Table 12.11 Binding energy for silicon/silicon fusion.

Fusion reaction components	Total BE (MeV)	SAM BE (MeV)	Difference BE (MeV)
Silicon-28	236.54	238.08	–1.54
Silicon-28	236.54	238.08	–1.54
Silicon-28 + Silicon-28	473.08	476.16	–3.08
Iron-56	492.26	491.73	0.53
Iron-56 – (Silicon-28 + Silicon-28)	19.18	15.57	–3.61

Even silicon/silicon fusion yields energy: 19.18 MeV per atom (Table 12.11).

What happens if we go above iron?

Table 12.12 Binding energy for iron/oxygen fusion.

Fusion reaction components	Total BE (MeV)	SAM BE (MeV)	Difference BE (MeV)
Iron-56	492.26	491.73	0.53
Oxygen-16	127.62	126.83	0.79
Iron-56 + Oxygen-16	619.88	618.56	1.32
Missing element 33/72	N/A	N/A	N/A
Germanium-72	628.69	638.58	–9.89
Germanium-72 – (Iron-56 + Oxygen-16)	8.81	20.02	–11.21

The reaction for iron/oxygen fusion is again complicated. The initial result actually—according to deuteron and single proton count—is a missing element with 33 deuterons and 72 protons (one deuteron split for the next carbon nuclet). It is unstable and rearranges to germanium-72 by splitting another deuteron for the next carbon nuclet and pulling in another outer electron as a quasi-inner electron. As a result, 8.81 MeV can be gained by this fusion process per atom, although not in a single step (Table 12.12).

Table 12.13 Binding energy for germanium/oxygen fusion.

Fusion reaction components	Total BE (MeV)	SAM BE (MeV)	Difference BE (MeV)
Germanium-72	628.69	638.58	–9.89
Oxygen-16	127.62	126.83	0.79

Fusion reaction components	Total BE (MeV)	SAM BE (MeV)	Difference BE (MeV)
Germanium-72 + Oxygen-16	756.31	765.41	−9.10
Strontium-88	768.47	785.43	−16.96
Strontium-88 – (Germanium-72 + Oxygen-16)	12.16	20.02	−7.86

For germanium/oxygen, 12.16 MeV can be gained by this fusion process per atom. Surprisingly for the Standard Model, fusion still yields energy above iron—not as much as below iron, but still enough. The energy balance is certainly not negative (Table 12.13).

Table 12.14 Binding energy for strontium/oxygen fusion.

Fusion reaction components	Total BE (MeV)	SAM BE (MeV)	Difference BE (MeV)
Strontium-88	768.47	785.43	−16.96
Oxygen-16	127.62	126.83	0.79
Strontium-88 + Oxygen-16	896.09	912.26	−16.17
Rhodium-104	891.17	932.28	−41.11
Rhodium-104 – (Strontium-88 + Oxygen-16)	−4.92	20.02	−24.94
Palladium-104	892.82	927.83	−35.01
Palladium-104 – (Strontium-88 + Oxygen-16)	−3.27	15.57	−18.84

Rhodium-104 is unstable and decays with a β– step to palladium-104. Now the energy balance is negative. The point where oxygen fusion no longer yields energy seems to be around strontium—not iron (Table 12.14).

For oxygen, it can be roughly located as follows: we look at the stored stress energy of the source as well as the target isotope. If the stress energy difference of the two is bigger than the SAM binding energy difference between the target and the source + oxygen (most of the time 20.02 MeV), then the fusion process is not endothermic. As for source and target isotopes below iron there is no relevant, stored stress energy and their energy difference is close to zero. Most of those reactions are therefore exothermic.

Table 12.15 Binding energy for iron/iron fusion.

Fusion reaction components	Total BE (MeV)	SAM BE (MeV)	Difference BE (MeV)
Iron-56	492.26	491.73	0.53
Iron-56	492.26	491.73	0.53
Iron-56 + Iron-56	984.52	983.46	1.06
Palladium-108	925.24	972.33	−47.09
Palladium-108 − (Iron-56 + Iron-56)	−59.28	−11.13	−48.15

Fusion of bigger chunks like iron-56 and iron-56 does not work energetically since it is not exothermic (Table 12.15).

Table 12.16 Binding energy for gold/helium fusion.

Fusion reaction components	Total BE (MeV)	SAM BE (MeV)	Difference BE (MeV)
Gold-197	1,559.38	1,771.1	−211.72
Helium-4	28.3	26.7	1.6
Gold-197 + Helium-4	1,587.68	1,797.8	−210.12
Mercury-201	1,587.41	1,811.15	−223.74
Mercury-201 − (Gold-197 + Helium-4)	−0.27	13.35	−13.62

The process of fusing gold-197 with helium-4 does barely not yield energy, this time leaving out the involved unstable missing element (Table 12.16).

Table 12.17 Binding energy for gold/deuterium fusion.

Fusion reaction components	Total BE (MeV)	SAM BE (MeV)	Difference BE (MeV)
Gold-197	1,559.38	1771.1	−211.72
Hydrogen-2	2.23	2.23	0.00
Gold-197 + Hydrogen-2	1,561.61	1,773.33	−211.72
Mercury-199	1,573.15	1,788.9	−215.75
Mercury-199 − (Gold-197 + Hydrogen-2)	11.54	15.57	−4.03

The process of fusing gold-197 with hydrogen-2 still yields energy: 11.54 MeV per atom. It comes down to what the smaller part can provide in addition to its own binding energy to the overall binding energy of the fused isotope. Helium-4 has comparatively high binding energy and provides only a small amount as additional binding energy when fused. The gain through deuterium is much greater (Table 12.17).

Table 12.18 Binding energy for polonium/helium fusion.

Fusion reaction components	Total BE (MeV)	SAM BE (MeV)	Difference BE (MeV)
Polonium-211	1,649.76	1900.15	−250.39
Helium-4	28.3	26.7	1.6
Polonium-211 + Helium-4	1,678.06	1,926.85	−248.79
Radon-215	1,669.22	1,926.85	−257.63
Radon-215 − (Polonium-211 + Helium-4)	−8.84	0.00	−8.84

Here we look again at the energy balance of the α decay of radon-215, but this time as α capture/absorption of polonium-211 (Table 12.18). Again, we see the 8.84 MeV as before (end of Section 7.3).

Table 12.19 Binding energy for plutonium/deuterium fusion.

Fusion reaction components	Total BE (MeV)	SAM BE (MeV)	Difference BE (MeV)
Plutonium-239	1,806.92	2,158.25	−351.33
Hydrogen-2	2.23	2.23	0.00
Plutonium-239 + Hydrogen-2	1,809.15	2,160.48	−351.33
Americium-241	1,817.93	2,176.05	−358.12
Americium-241 − (Plutonium-239 + Hydrogen-2)	8.78	15.57	−6.79

Even a fusion of plutonium-239 with deuterium still yields energy: 8.78 MeV (Table 12.19). In essence: we run out of possible structure before we run out of binding energy.

> It is very clear that the average binding energy graph (Fig. 7.5), which is used to argue about the role of the area around iron/nickel/zinc as the energetic end point of fusion, is misleading. We found that fusion can always happen. However, depending on what is fused, the point where the process changes from exothermic to endothermic varies. Also the result might not be stable.

12.3 HOW REALISTIC ARE FUSION SCENARIOS?

We looked extensively at oxygen fusion in the energy calculation examples above. However, fusion is not as simple as adding deuterons, single protons, and PEPs together and determine the summary outcome. Structure plays a major role during fusion. If a big resettling step is required to create the structure of the fused new element, then chances are high the fusion step would fail, resulting in smaller elements and a few leftover protons and PEPs.

12.4 WHERE DOES FUSION HAPPEN?

Usually, the fusion of elements is connected to the genesis of stars, and fusion in the Standard Model is considered to happen inside stars. Fusion reactions that produce elements heavier than iron are endothermic according to the Standard Model and are unable to generate the energy required to maintain stable fusion inside the star. This statement is actually a false notion. Fusion after iron can still be an energy-yielding process. However, on average it is less desirable. Furthermore, we must consider that stars are, for the bigger part (if not all), an electrical phenomenon (Section 10.2). We can only see the surface of a star, we cannot look inside. There *are* nuclear reactions happening on the surface of a star, but is it really fusion? It might very well be fission (Chapter 13) or a combination of both. The sometimes difficult to accept reality is that we still know next to nothing about the inner parts of the Sun. For obvious reasons we cannot simply do observations of the interior. However, if, say, we consider sunspots to be a window into the interior of a star, then we see blackness, which might be either something cold or we see nothing. According to the Standard Model we would expect to see a hot fusion furnace. Some theories and models exist that show a very different picture for the makeup of the Sun, from a hollow plasmoid [Thornhill 2010], to a full-fledged planet-like inner part [Thornhill 2021] or a metal-hydrogen lattice [Robitaille 2011]. Some of these options more or less exclude interior fusion.

So where is fusion happening in nature on a large scale if not in stars? It could happen in planets—at least on large scales at times of immense electrical stress in the solar environment. We think that the heavier elements are in fact predominantly created on planets during electrically violent phases of a planet's life. We will look further into this topic in Section 15.4.

CHAPTER 13

Fission

Nuclear fission is a nuclear reaction or a radioactive decay process in which the nucleus of an atom splits into two or more smaller, lighter nuclei. The fission process releases a very large amount of energy.

13.1 A LITTLE BIT OF HISTORY

The nuclear fission of heavy elements was discovered in 1938 by Otto Hahn. This discovery process was not as straightforward as it may seem today. Initially, the expectation of Hahn, Meitner, Strassmann, and others was that shooting "neutrons" onto uranium would yield transuranium elements. At that time, the existence of the actinides was not yet established, and uranium was believed to be an element similar to tungsten. It followed that the first transuranic elements would be similar to rhenium and the platinoids. Hahn et al. established the presence of multiple isotopes of at least four such elements and identified them as elements with atomic numbers 93 through 96, which was an error—but that was not clear to them. In late 1938 they found evidence of isotopes of an alkaline-earth metal in their sample. Finding an alkaline-earth metal was problematic because it did not logically fit with the other elements found thus far. Hahn initially suspected it to be radium, produced by splitting off two α particles from the uranium nucleus. At the time, the scientific consensus was that even splitting off two α particles via this process was unlikely. The idea of turning uranium into barium (by removing around 100 nucleons) was seen as preposterous. Further refinements of the technique, leading to the decisive experiment on 16/17 December 1938, produced the same puzzling result: the isotopes found behaved like barium. By January 1939, Hahn was sufficiently convinced that the formation of light elements was occurring in his setup. He published a revision of his article, essentially retracting former claims of observing transuranic elements and neighbors of uranium and concluding instead that he was seeing light platinoids, barium, lanthanum, and cerium.

13.2 PROCESS DETAILS

The two (or more) nuclei produced during fission are most often of comparable but slightly varying sizes, typically with a mass ratio of the fission products of about 3 to 2, for common fissile isotopes. Most fission events are binary (producing two fragments), but occasionally three fragments are produced.

Fission is usually induced by neutron bombardment of a heavy isotope. Isotopes undergo fission when struck by neutrons, and in turn emit neutrons when they break apart. This makes a self-sustaining nuclear chain reaction possible, releasing energy at a controlled rate in a nuclear reactor or at a very rapid, uncontrolled rate in a nuclear weapon. A natural form of spontaneous radioactive decay is also referred to as fission and occurs especially in very heavy isotopes. Spontaneous fission of uranium was discovered in 1940.

Fissile material can sustain a chain reaction with thermal neutrons by definition. A fissionable nucleus that can be induced to fission with thermal neutrons with a high probability is referred to as "fissile." Fissionable materials include those that can undergo fission only with high-energy neutrons—like uranium-238. Fissile materials are therefore a subset of fissionable materials.

13.3 THE ASYMMETRIC BREAKUP OF THE NUCLEUS

The composition of fission products is unpredictable when adopting "currently accepted theory." An even split would be expected by standard models of the nucleus, but instead two peaks are seen, roughly in the ratio of 3:2 (Fig. 13.1).

The *initial fission products* before PEP release are exactly mirrored. The data needs a closer look. It is collected from Cook [2010, 155–157] (Tables 13.1–13.3).

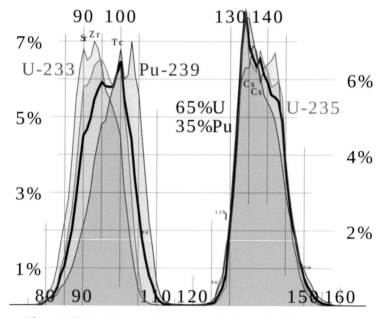

Figure 13.1 Fission product composition for various heavy isotopes.
[Wikipedia 2021/Nuclear_fission_product]

Table 13.1 Initial fission results for uranium-233 + PEP → uranium-234 before PEP release.

Heavy		Light	
# Nucleons	# Outer electrons	# Nucleons	# Outer electrons
143	56 (Barium)	91	36 (Krypton)
139	54 (Xenon)	95	38 (Strontium)
135	52 (Tellurium)	99	40 (Zirconium)

Table 13.2 Initial fission results for uranium-235 + PEP → uranium-236 before PEP release.

Heavy		Light	
# Nucleons	# Outer electrons	# Nucleons	# Outer electrons
144	56 (Barium)	92	36 (Krypton)
139	54 (Xenon)	97	38 (Strontium)
134	52 (Tellurium)	102	40 (Zirconium)

Table 13.3 Initial fission results for plutonium-239 + PEP → plutonium-240 before PEP release.

Heavy		Light	
# Nucleons	# Outer electrons	# Nucleons	# Outer electrons
143	56 (Barium)	97	38 (Strontium)
138	54 (Xenon)	102	40 (Zirconium)
135	52 (Tellurium)	105	42 (Molybdenum)

It appears that the lighter fission product grows as the nucleon number grows (Fig. 13.2), while the nucleon number of the heavier part stays more or less the same.

We first look in more detail at uranium-235 and a neutron—the intermediate result would be the even more unstable uranium-236, which then would fission into, for example, krypton-92 and barium-141 and three neutrons, as well as gamma rays (Table 13.2). All the fission products would have high kinetic energy. Why we see this distribution no standard model in use can explain. Let's have a look at those nuclei in SAM. We start with uranium-235 (Fig. 13.3).

Figure 13.4 shows the side view of uranium-235. The two branches on top of the nucleus or in the middle of the picture nearly touch each other—this is clearly visible.

Uranium-236, at least if given a thermal PEP, would look like Figure 13.5. Why it is not fissile by thermal "neutrons" is now visibly apparent: all 9 spots for thermal "neutrons" are taken. In ~15% of the cases of supplying a PEP to uranium-235 it just becomes uranium-236 and releases a gamma ray.

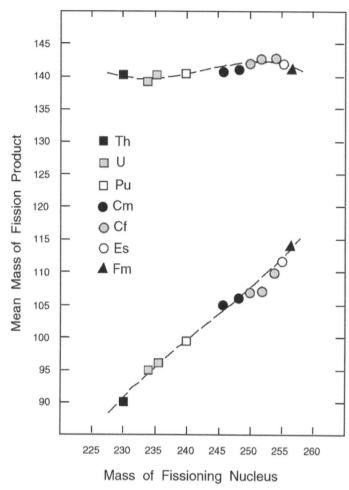

Figure 13.2 A summary of light and heavy fragment masses.
[Cook 2010, 153]

However, in ~85% of cases this incoming thermal "neutron" hitting uranium-235, creating uranium-236 in the process, does something else: it attaches to one of the nearly colliding branches in the receptive spot and, in combination with the kinetic energy of the impact, causes the two branches to fuse (with the help of the quasi-inner electron between the branches) and to simultaneously break off at the final endings where they connect to the backbone. Barium-141 as the heavier fission product is shown in Figure 13.6. And krypton-92, as the fused result of the two branches, is shown in Figure 13.7.

The broken off branches carry 38 and 54 protons, respectively (Fig. 13.8), which represent the 92 protons of krypton-92. When the branches fuse (which the Atom-Viewer tool currently cannot show), they rearrange protons and inner electrons to the

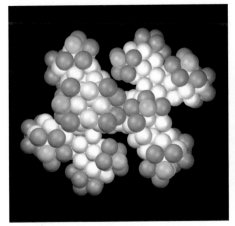

Figure 13.3 Uranium-235, *top view.*

Figure 13.4 Uranium-235, *side view.*

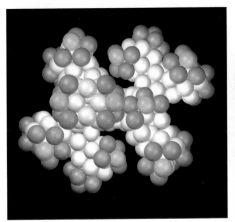

Figure 13.5 Uranium-236, not receptive to PEPs.

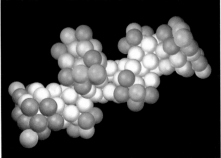

Figure 13.6 Barium-141, bigger part of the fission process.

new isotope (here krypton-92) which might be stable or—in this case—unstable. This rearrangement causes a lot of radiation.

The branches fuse together with the help of quasi-inner electrons that reside between the branches and pull them together. As a result, the branches (now fused) break at the "bottom of the branch" and collectively depart the nucleus.

From the remaining core backbone of 144 protons and a number of inner electrons, 3 PEPs are ejected, creating after rearrangement the bigger fission product (here barium-141) which again might be stable or—in this case—unstable.

It is possible that the two branches that have broken off do not fuse permanently and fission again or never fuse at all. In this case we get ternary fission.

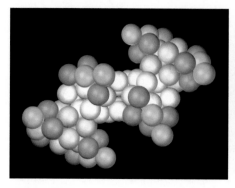

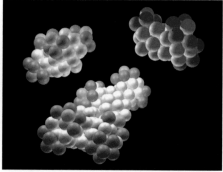

Figure 13.7 Krypton-92, smaller part of the fission process.

Figure 13.8 Uranium-236 with two branches broken off (54 and 38 protons).

This is just one fission example, but the message is clear: in this example SAM neatly explains the asymmetric breakup of nuclear fission based on the structure of the nucleus. And something else is becoming clear: the fission of heavier isotopes is not just creating smaller parts, it is also about fusion, those parts broken off partially recombining—with the help of quasi-inner electrons between branches. Not only does it become clear that the heavier fragment represents the backbone and the lighter fragment represents the fused branches where most of the growth happens (Fig. 13.2), but the numbers are an excellent fit as well, at least in this example. Figure 13.2 shows that this 141/92/3 fission (p. 178) is just one outcome of several possible for uranium-235—Table 13.2 shows only the peak examples. We argue that we might not always get a clean breaking off of branches, as shown in Figure 13.8, since this depends on the angle of impact of the PEP.

In the case of plutonium-239/240 the SAM numbers are currently off by one proton. We would expect a clean cut at 144/96—not at 143/97 (as given in Table 13.3). With uranium-233/234 the SAM numbers are currently also off by one proton. We would expect a clean cut at 144/90—not at 143/91 (as given in Table 13.1). When branches break off, the question that always arises is where the shared proton goes. So, the literature numbers can be achieved with SAM. We just don't know yet why it goes in one direction with the first case and in the other direction with the last two cases.

> SAM explains the asymmetric breakup of the nucleus during a fission event through the structure of the nucleus.

13.4 ENERGY RELEASE DURING FISSION

This is what *Wikipedia* has to say about the energy release topic:

Typical fission events release about two hundred million eV (200 MeV) of energy, the equivalent of roughly >2 trillion Kelvin, for each fission event. The exact isotope which is fissioned, and whether or not it is fissionable or fissile, has only a small impact on the amount of energy released. This can be easily seen by examining the curve of binding energy (image below) [see Fig. 7.5 in this book], and noting that the average binding energy of the actinide nuclides beginning with uranium is around 7.6 MeV per nucleon. Looking further left on the curve of binding energy, where the fission products cluster, it is easily observed that the binding energy of the fission products tends to center around 8.5 MeV per nucleon. Thus, in any fission event of an isotope in the actinide's range of mass, roughly 0.9 MeV is released per nucleon of the starting element. The fission of U235 by a slow neutron yields nearly identical energy to the fission of U238 by a fast neutron. This energy release profile holds true for thorium and the various minor actinides as well. [Wikipedia 2021/Nuclear_fission]

The calculation being done is very simple: 0.9 MeV times 235 nucleons equals 211.5 MeV. Now let's have a look at the binding energy values for our prime example of uranium-236 fissioning to barium-141 and krypton-92. For the calculation, we handle this exactly like the fusion calculations (Section 12.2) and reverse the results (Table 13.4).

Table 13.4 Binding energy for barium-141/krypton-92/3 "neutron" fusion.

Fusion reaction components	**Total BE (MeV)**	**SAM BE (MeV)**	**Difference BE (MeV)**
Barium-141	1,173.98	1,259.35	−85.37
Krypton-92	783.17	812.13	−28.96
3 "neutrons"	2.35	20.03	−17.68
Barium-141 + Krypton-92 + 3 "neutrons"	1,959.50	2,091.51	−132.01
Uranium-236	1,790.41	2,133.78	−343.37
Uranium-236 − (Barium-141 + Krypton -92 + 3 "neutrons")	−169.09	42.27	−211.36

Fusing barium and krypton together with 3 "neutrons" into uranium-236 is a highly endothermic process since 169.09 MeV needs to be spent. The difference in stress energy to be stored in the nucleus is even higher: 211.36 MeV. Fortunately, 42.27 MeV can be freed once fusion is complete.

As a fission process, the amount of binding energy equal to 169.09 MeV is released after the first fission step. The full 211.36 MeV of stress energy released cannot be used since a portion of it goes into the structure of the nucleus (42.27 MeV). More decay steps follow, bringing the amount of released energy to roughly 200 MeV, which is the literature value commonly given for uranium fission.

> The "stress energy" represents a hidden source of energy in the nucleus that can be tapped into through fission.

13.5 STRUCTURAL THOUGHTS AND OTHER FISSION RELEASES

The barium area around 120–140 protons is where the middle branches are least developed. Disintegrating heavy atoms by breaking off middle branches would indeed leave mostly elements in this range as the heavier part. The range from tin-114 to barium-130 is like a boundary (i.e., after which middle branches need to be made). If we start around the element tin in the Atom-Viewer moving upward, then the barium boundary is clearly visible. After barium the rare-earth elements start and this is where the two new branches need to be created. These branches form the "collision points" which are used in the fissioning of heavier elements. The third collision point is formed around lead. This one is a little closer to the middle branches and more "edgy" for here the real unstable elements start. It seems that the existence of a collision point is essential for fission that is induced by PEPs.

We must also consider what happens to some of the quasi-inner electrons during fission. Additional quasi-inner electrons are located between the branches. If a nucleus fissions then these electrons will be set free. A fission process should show measurable amounts of beta radiation, and indeed it does:

> *A second blackout effect is caused by the emission of beta particles from the fission products. These can travel long distances, following the Earth's magnetic field lines. When they reach the upper atmosphere they cause ionization similar to the fireball, but over a wider area. Calculations demonstrate that one megaton of fission, typical of a two megaton H-bomb, will create enough beta radiation to blackout an area 400 kilometers (250 mi) across for five minutes. Careful selection of the burst altitudes and locations can produce an extremely effective radar-blanking effect.* [Wikipedia 2021/Effects_of_nuclear_explosions#Electromagnetic_pulse]

13.6 THE THORIUM CYCLE

Thorium, due to its abundance, is a potential nuclear fuel. The thorium cycle creates uranium-233 in a first step. This is the real fissile material that yields the sought-after energy. The cycle occurs because uranium-233 fission yields some "neutrons" (PEPs) that in turn can enrich more thorium-232 into thorium-233, which then decays through protractinium-233 to uranium-233 (Fig. 13.9).

The reaction takes place in a molten salt reactor which is inherently safe by design. The fission products are still a problem in the sense that they tend to be unstable isotopes decaying into stable products. The advantage is that they are contained in the reactor. Should we be able to solve the issue of these accumulating radioactive isotopes (fission products) then we would be able to make the whole cycle very clean and safe for the environment. The abundance of thorium in very accessible beach sand deposits makes it even more attractive.

This type of nuclear energy would be in contrast to current water-pressurized systems in that they could not explode or melt down. Instead, they would stop automatically the moment the heat supply was removed. If the cooling of the whole system failed, then a

External PEP source as starter

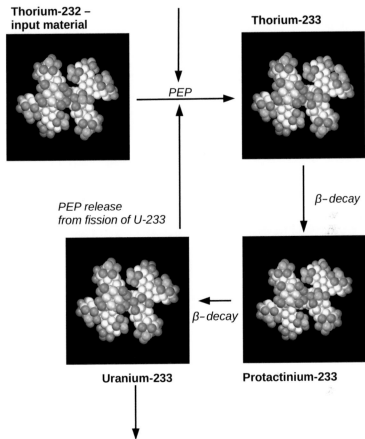

**Thorium-232 –
input material**

PEP

Thorium-233

β–decay

*PEP release
from fission of U-233*

β–decay

Uranium-233

Protactinium-233

Fission of U-233 into fission products such as Sr, Zr,
Sn, I, Cs, La, Ce + 2/3 neutrons – **output material**

Figure 13.9 Schematic thorium cycle. We start with PEP capture of thorium-232, resulting in thorium-233, which then reacts with a β– decay step converting to protactinium-233, which again utilizes a β– decay step to go to uranium-233, which fissions and creates "neutrons" in the process which again can be captured by thorium-232, which starts the cycle again.

safeplug simply melts and the molten salt falls into a tank, thereby immediately stopping the reaction.

Thorium itself is not a fissile material and thus cannot undergo fission to produce energy. Instead, it must be transmuted to uranium-233 in a reactor fueled by other fissile materials. The first two stages, natural uranium-fueled heavy water reactors and plutonium-fueled fast breeder reactors, are intended to generate sufficient fissile material from limited uranium resources such that the vast thorium reserves can be fully utilized in the third stage of thermal breeder reactors.

At Oak Ridge National Laboratory in the 1960s, the Molten Salt Reactor Experiment used uranium-233 as the fissile fuel in an experiment to demonstrate a part of the molten salt breeder reactor that was designed to operate on the thorium fuel cycle. Molten salt reactor (MSR) experiments assessed thorium's feasibility, using thorium fluoride dissolved in a molten salt fluid to eliminate the need to fabricate fuel elements. The MSR program was defunded in 1976 after its patron Alvin Weinberg was fired.

CHAPTER 14

Lessons learned

We have learned a lot about nuclear reactions in the last several chapters. Let's take a moment to summarize what was discussed:

- The buildup of the nuclei of elements follows a plan directed by at least four organizational growth patterns. There is a clearcut correlation between structure and stability on several layers of these growth patterns. We could even add a 5th-order organizational pattern: the nucleus wants to be in a noble configuration. This is a clear result of the nuclei buildup we see in the periodic table of elements (PTE).
- As mentioned before there are three types of electrons at play: outer, which the Standard Model acknowledges; inner, which keep the proton structure of the nucleus together like glue; and quasi-inner, which reside just above the topside of the nucleus between the endings of branches and between the branches themselves.
- Those quasi-inner electrons were initially introduced as our solution for the conundrum of the "neutron"/proton ratio of the heavier nuclei. We later found that they play an important role in the stability of the nuclei as they tend to degrade it by pulling branches together, resulting in "stress energy" stored in the nucleus.
- There is "stress energy" stored within the nuclei, which shows up as the deviation between the SAM binding energy data and the published binding energy data. This stress energy is the energy that can be (at least partially) released through the fission of a nucleus.
- The quasi-inner electrons can also cause fusion by pulling branches together. As a consequence, we see the numbers of the lighter fission results related to the asymmetric breakup of the nucleus.
- The asymmetric breakup of the heavier nuclei is explained by SAM. It follows from the elongated backbone structure, which—once the middle branches come off—becomes the heavier fission product, while the branches fuse and provide the lighter fission product. The numbers match remarkably well.
- The nucleus provides connection spots based on the specific structure for each element. An outer electron from one atom connects to one of those spots on another nucleus and vice versa. The (more positive) spots can be used to make chemical bonds but also nuclear bonds. Those are also the spots where additional PEPs can latch onto the nucleus with a strong connection. In turn, those PEPs (now a more negative spot relative to the surroundings) provide connection spots for additional

protons (proton capture) to create the next deuteron and cause transition to the next element, after resettling,.

- Proton capture has emerged as the prime candidate for causing immediate nuclear reactions. However, there are preconditions to be met. The isotope in question needs to be receptive to protons, it needs to have a boron ending or some five-endings. Also, the environment needs to be electrically active, that is, a plasma or an electrical discharge needs to be present.
- A new numbering scheme for the PTE was introduced. It is no longer based on the number of protons/outer electrons as in the Standard Model, but on the number of deuterons and single protons (at least those single protons not involved in pulling in a quasi-inner electron into the nucleus). It opened up space in the PTE for missing elements. The symmetries showing up in this new PTE (Appendix B) are fascinating and by no means complete or completely understood.

14.1 SOME VERY BASIC QUESTIONS

In her paper "The Electromagnetic Considerations of the Nuclear Force" Nancy L. Bowen asks some important questions. She states:

The following is a partial list of some very basic questions that cannot be answered by the current models. [Bowen 2018]

After giving the list, she states:

For most of these questions, the current models cannot offer any theoretical answer, other than to say, 'because experimental energy levels allow the transition.' This response is not a valid theoretical answer, and is merely an experimental observation in support of the laws of thermodynamics. The current nuclear models themselves offer no theoretical explanation. [Bowen 2018]

Here are Nancy Bowen's questions:

1. Why do certain medium-sized isotopes, such as ^{150}Dy, exhibit alpha radiation, and yet certain larger isotopes, such as ^{208}Pb, do not?
2. Why is ^{8}Be so extremely unstable, and why does it exhibit alpha radiation?
3. Considering the daughter nuclides after the fission of ^{235}U, why is there an extreme double-humped curve?
4. Why do certain light isotopes (such as ^{4}He, ^{8}Be, ^{12}C, ^{16}O, ^{20}Ne, etc.) have a spike in their binding energy per nucleon?
5. Why does ^{4}He have a zero thermal nuclear cross-section?
6. Why is ^{5}He so extremely unstable?
7. Why are the magnitudes of quadrupole moments so much larger than any model predicts?
8. What causes a nucleus to eject a neutron (aka neutron decay) in certain nuclides?

9 What causes proton decay in certain nuclides?
10 What causes alpha decay in certain nuclides?
11 What causes spontaneous fission in certain nuclides?
12 What causes beta-delayed neutron decay in certain nuclides?
13 What causes beta-delayed proton decay in certain nuclides?
14 What causes beta-delayed alpha decay in certain nuclides?
15 What causes beta-delayed spontaneous fission in certain nuclides?

If you have read this book carefully, you will be able to answer most of these questions. See Chapter 9 and Sections 1.4, 4.1, 11.3, 11.5, 11.8, and 13.3 for the answers. If anything, this makes clear what SAM brings to the table.

14.2 ARE THERE MISSING ELEMENTS?

Throughout the book we found increasing evidence for missing, currently unknown elements. Switching to a deuteron count for elements then made it clear that they are in fact a crucial part of the SAM model. The first hint were the noble gases: argon is composed of two completely capped carbon nuclets and krypton, the next known noble element, has five completely capped carbon nuclets. The reason, of course, is the parallel application of three cycles-of-eight intertwined. However, one wonders whether there is a noble element with three (Fig. 14.1) or four (Fig. 14.2) completely capped carbon nuclets? In theory this should be the case, but all element numbers in this number range are used up, at least if we stay with the original Standard Model definition of an element and its number—defined by the number of outer electrons. We created a new system based on the deuteron plus single proton count, which clearly showed gaps in the current lineup. This new system identified that most cases of double β decays are related to missing—unstable—elements and thus are explainable.

For a more systematic approach we can look at all possible endings and nuclet permutations (with left and right interchangeable) on top of a carbon nuclet (Table 14.1).

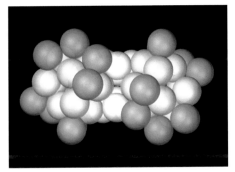

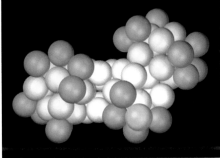

Figure 14.1 Missing noble gas with 50 protons.

Figure 14.2 Missing noble gas with 66 protons.

Table 14.1 Theoretical ending/nuclet permutation on top of a base carbon.

Left	Right	Proton setup	# Protons including base carbon (12)	# Deuterons including base carbon (6)	Isotope
			Comment		
None	Two-ending	0 + 2	14	7	Nitrogen-14
Known element on top of the initial carbon nuclet, otherwise not to be found again as stable element—at least as a single nuclet.					
None	Four-ending	0 + 4	16	8	Oxygen-16
Known element on top of the initial carbon nuclet.					
None	Five-ending	0 + 5	17	8	Unknown-17
This was an option for oxygen-17, but the binding energy says otherwise.					
None	Lithium nuclet	0 + 6	18	9	Unknown-18
Unbalanced.					
None	Beryllium ending	0 + 8	20	10	Unknown-20
Unbalanced.					
None	Boron ending	0 + 10	22	11	Unknown-22
Unbalanced.					
None	Carbon nuclet	0 + 11	23	11 + 0.5	Unknown-23
Unbalanced.					
Two-ending	Two-ending	2 + 2	16	8	Unknown-16
Before the first two-ending can appear on the second growth point, we need at least a five-ending on the other growth point.					
Two-ending	Four-ending	2 + 4	18	9	Unknown-18
Before the first two-ending can appear on the second growth point, we need at least a five-ending on the other growth point.					
Two-ending	Five-ending	2 + 5	19	9	Fluorine-19

Left	Right	Proton setup	# Protons including base carbon (12)	# Deuterons including base carbon (6)	Isotope
Comment					
Two-ending	Lithium nuclet	2 + 6	20	10	Unknown-20
Unbalanced.					
Two-ending	Beryllium ending	2 + 8	22	11	Unknown-22
Unbalanced.					
Two-ending	Boron ending	2 + 10	24	12	Unknown-24
Unbalanced.					
Two-ending	Carbon nuclet	2 + 11	25	12 + 0.5	Unknown-25
Unbalanced.					
Four-ending	Four-ending	4 + 4	20	10	Neon-20
Four-ending	Five-ending	4 + 5	21	10	Neon-21
Four-ending	Lithium nuclet	4 + 6	22	11	Sodium-22
Unbalanced. Unstable isotope.					
Four-ending	Beryllium ending	4 + 8	24	12	Unknown-24
Unbalanced. This was a possible option for magnesium, more in line with the beryllium group, but we went for the double lithium nuclet instead.					
Four-ending	Boron ending	4 + 10	26	13	Unknown-26
Unbalanced.					

Left	Right	Proton setup	# Protons including base carbon (12)	# Deuterons including base carbon (6)	Isotope
			Comment		
Four-ending	Carbon nuclet	4 + 11	27	13 + 0.5	Unknown-27
Unbalanced.					
Five-ending	Five-ending	5 + 5	22	10	Neon-22
Five-ending	Lithium nuclet	5 + 6	23	11	Sodium-23
Five-ending	Beryllium ending	5 + 8	25	12	Unknown-25
Not densely packed.					
Five-ending	Boron ending	5 + 10	27	13	Unknown-27
Not densely packed. Another possible option for aluminum.					
Five-ending	Carbon nuclet	5 + 11	28	13 + 0.5	Silicon-28
Lithium nuclet	Lithium nuclet	6 + 6	24	12	Magnesium-24
Lithium nuclet	Beryllium ending	6 + 8	26	13	Unknown-26
Not densely packed.					
Lithium nuclet	Boron ending	6 + 10	28	14	Unknown-28
Not densely packed.					
Lithium nuclet	Carbon nuclet	6 + 11	29	14 + 0.5	Unknown-29

Left	Right	Proton setup	# Protons including base carbon (12)	# Deuterons including base carbon (6)	Isotope
				Comment	
Unstable, the quasi-inner electron will break it apart.					
Beryllium ending	Beryllium ending	8 + 8	28	14	Unknown-28
There is a more densely packed option (silicon-28)					
Beryllium ending	Boron ending	8 + 10	30	15	Unknown-30
Probably unstable.					
Beryllium ending	Carbon nuclet	8 + 11	31	15	Phosphorus-31
Boron ending	Boron ending	10 + 10	32	16	Unknown-32
There is a more densely packed option (sulfur-32)					
Boron ending	Carbon nuclet	10 + 11	33	16 + 0.5	Unknown-33
Probably stable.					
Carbon nuclet	Carbon nuclet	11 + 11	34	16 + 0.5 + 0.5	Unknown-34
Probably stable.					

These are 35 possible combinations, only 31 when counting deuterons and structure. Only 8 of these options are actual known elements (*light gray*) and 3 are isotopes of known elements (*dark gray*). The rest is unknown.

This raises the issue of the underlying rationale for the choice of a particular configuration because often there are several options available. Here the pillars of observation (Section 5.7.1) play a significant role in determining the most likely configuration for an element:

- PEP/proton ratio
- known nuclear reactions
- isotopes and stability

- valence/oxidation state
- abundance data.

Armed with these tools and information, we chose the best fit. However, that does not mean that those "roads not taken" cannot be used. In most cases the elements based on these structures will very likely be unstable. However, there are cases where we do not see a reason for instability.

One example: when we look at the SAM nuclear PTE (Appendix B) we see that the elements which would be similar to oxygen only two steps before a noble gas state are missing below sulfur. Those elements could be more oxygen-like than the elements currently put into the oxygen group.

Looking at the mentioned trends in abundance (Section 2.20.2) of the elements of the oxygen group, it is suspected that these elements may simply not be created or exist only in very low abundance. The stability of these configurations may also be a factor to explore, which would mean that if, say, missing element 57, which would also be similar to oxygen, is created and assumed to be unstable, it could decay to an isotope of tellurium. Examining the structure of the elements in the oxygen group reveals that the heavier they become, the more diverse the active endings become, yielding multiple inherited chemical properties, hence the element is less like oxygen. The idea that we see these shifts in the PTE (representing the groups ordered according to chemical properties) is considered another research topic.

## CHAPTER 15

Transmutation

After an introduction, we will first explore the various "dimensions" of the term *transmutation* before returning in more detail to transmutation experiments that have been done over the last 30 years. These topics include history, biology, and geology. The acceptance of transmutations on Earth will have a great impact on these fields in the future. *In this book we only want to touch upon what the consequences of the introduction of transmutation into these fields are.*

However, there is a general belief going back almost a century that transmutations take place in stars and in supernovae [Wikipedia 2021/Supernova_nucleosynthesis], and that most elements on Earth were created at least 4.5 billion years ago. Most physicists reject the idea of laboratory transmutations based on this consensus point of view and assume that the reactions needed to modify the nucleus can only occur in energetic processes such as theorized in the Big Bang, stellar hot fusion, and in powerful manmade reactors on Earth. Based on SAM, we disagree with this consensus belief. First, let us define transmutation.

> **Definition**
> Transmutation is the in situ change from one element to another, facilitated by the structure of the nucleus, the outer electron configuration, the type of nuclear reaction, and the environment in which the nucleus exists.

The energies required to induce transmutation vary based on the nuclei involved and the environment. When the energy input to induce a nuclear reaction is low compared with the energy output, we would like to use the term "Low Energy (induced) Nuclear Reaction" (LENR). The term "induced" in parentheses makes clear we are talking about low *input* energies—not low output energies. We will take the term LENR literally here and use it just for what the words mean, initially ignoring a more specific meaning that has been attached to the term in the last several years.

15.1 TRANSMUTATION AND LOW-ENERGY NUCLEAR REACTIONS

In this context nearly all the nuclear reactions we have talked about in this book (Table 15.1) have the potential to be low-energy nuclear reactions. Most nuclear reactions result in the transmutation of one element into another.

Table 15.1 Nuclear reactions and transmutation.

Nuclear reaction	Transmutation
β– decay	✓
β+ decay	✓
α decay	✓
α capture	✓
+PEP	
–PEP	
Proton capture	✓
Proton emission	✓
Fusion	✓
Fission	✓

Changes in the PEP count of a nucleus will usually result in a change into another isotope of an element. Although this is a change in the nucleus, it is not a change directly to a different element. All other nuclear reactions we discussed in this book result in direct transmutations (Table 15.2). PEP capture can, of course, cause a fission reaction with isotopes like uranium-235 which then results in a transmutation step. And adding a PEP can, of course, cause a β– decay as a consequence.

Some of these nuclear reactions happen "spontaneously," that is, they require very low input energy. Even uranium fission can happen spontaneously and the technical process used in nuclear power plants is not very energetic on the input side—just offer a few thermal PEPs—especially if compared with the fission output of 170 to 200 MeV per atom. This considerably high amount of energy output is based on the fission process tapping into the "stress energy" stored in the nucleus (Section 13.4). However, this is specific to the fission of heavy elements. Another prime candidate for transmutation is proton capture (Section 11.11). While PEP capture followed by a β– decay shows the same end result, it very often involves a resettling of the

Table 15.2 Classification of reactions.

Transmutation									
Fusion			**Simple nuclear reactions**					**Fission**	
				Decay					
Bigger nuclei	α capture	Deuteron capture	Proton capture	β– decay	β+ decay	Proton emission	Deuteron emission	α decay	Bigger nuclei

nucleus (which means potentially bio-hazardous radiation being emitted). A proton gently delivered to the right place of the nucleus can go without resettling (and without hazardous radiation).

15.2 TRANSMUTATION DURING HISTORICAL TIMES

In historical times the process of changing one element into another was called *alchemy*. Whether it was ever successful with the methods and tools available during those times remains an open question. We cannot exclude this from being the case because a lot of information has been lost over time. The term alchemy was also used for practical knowledge about creating metals and alloys, which in turn were transformed into weapons, tools, and jewelry. One of the reasons alchemy is such a controversial topic is that it makes no distinction between chemical effects and the idea of transmutation, which belongs more in the traditional domain of physics.

Many metals need to be mined from minerals such as PbS, SnO, or CuO. The raw metal then needs to be released from the mined material. A crystal with distinct colors and form would "transmute" into a dense grayish shiny metal. This process would look like transmutation to a casual observer, but it is of course a chemical process. There are also some naturally occurring pure elements such as gold. Pure sulfur can be found near volcanoes. The art of handling those materials and how to combine them to create desired effects was one of the big secrets of ancient times. Those are now common knowledge, but the in situ transmutation of elements has eluded us. Many people think this is not possible on Earth. We disagree—not only do we think it can be done in a laboratory but we also think it has happened in nature all around us with some transmutation processes being part of our daily lives.

15.3 TRANSMUTATION IN NATURE—BIOLOGY

C. Louis Kervran, a French scientist, published books and papers about the possibility of biological transmutations. In his book *Biological Transmutations* [Kervran 1980] many examples are given about specific nuclear reactions. A selection follows:

- Potassium + Hydrogen → Calcium
- Magnesium + Oxygen → Calcium
- Carbon + Oxygen → Silicon.

Every cell has what is known as mitochondria. They are considered to be the power plants of a cell—based on chemical processes. However, they are also suspected to be a potential place where charge separation of double-layers occurs (Section 10.2).

We have already pointed out how, with SAM, the absorption of hydrogen (a proton plus an outer electron) is explained as proton capture. The example of chickens converting potassium to calcium when given calcium-depleted food is a well-known example of transmutation happening inside a living being. The energy released ranges between 8 MeV and 0.5 MeV per atom depending on the source isotope.

The energy released when looking strictly at the binding energy differences of source and target element usually ranges in the area of several kiloelectron volts (KeV) to multiple megaelectron volts (MeV) per atom. One MeV is equal to 0.0000000000001602 joule. A joule is equal to the energy transferred to an object when a force of one newton acts on that object in the direction of the force's motion through a distance of one meter. One joule can also be understood as the typical energy released as heat by a person at rest every 17 ms. One joule is also the heat required to raise the temperature of 1 g of water by 0.24°C. A precursor for biological transmutations would therefore be the precise control of the transmutation rate possibly down to single atoms, otherwise too much energy could be created at once, killing the organism in the process. Keeping that in mind, the idea of transmutations in biology is not that strange after all.

Kervran also theorizes—supported by observations and measurements—on how the continental crust is too rich in calcium to be explained according to current theories. He points out that there is no conventional way to explain the accumulation of calcium through organisms based on chemical processes. We will discuss this theory in Section 15.5 when we look in more detail at element abundances. The fusion of larger elements as a process happening in nature, such as carbon and oxygen creating silicon, has been almost completely ignored as a possibility in consensus science. These reactions are not studied, nor has any attempt been made to study them.

15.4 TRANSMUTATION IN NATURE—GEOLOGY

The typical elements we find in nature are mostly located in specific areas and strata. Vein-like structures in geology are the main source for many of the elements/metals we extract from the Earth in mines. A question that one could ask, is: When these vein-like structures were created, were there instant "transmutations" of the elements taking place? What process caused those veins?

The proposed answers to these questions link directly back to the theory of the "Electric Universe," a universe which is formed by electric currents and resulting forces, rather than gravity (Sections 10.2 and 10.3). Bruce Leybourne published a paper [Leybourne 2017, 340] about the relationships between lightning, gravity, and earthquakes. In this paper the Earth is shown as a "generator." When this system is under stress due to external factors (perhaps large solar flares?), it would show enormous electrical activity and induce electric arcing that is of a very large magnitude. Are all these vein-like structures a result of primordial arcing on a higher, more intense level than seen today? When such a Lichtenberg pattern occurs not in the sky but underground, we imagine that these pathways would be more or less instantly vaporized and ionized, while smaller fragments (breccia) and all gases and plasma created would be under huge pressure, trying to find a way out. Under these circumstances of big instant pressure and electric forces, combined with a plasma, nature would be able to perform nuclear reactions as described in this book. This completely depends on the elements and isotopes that are present and many more physical circumstances. This notion is also quite feasible since even normal lightning has been shown to produce gamma rays and "neutrons" [Gurevich et al. 2012]. We therefore hypothesize that next to the chemical sorting effects from electrical discharges, heating, and pressurizing, fusion and fission also take place (i.e., transmutation). If that is the case, then there should be a correlation between the layers of the Earth that were pierced through with their chemical composition and the elements that can be found in the veins. This is a topic we want to go into in much more detail in a future book. For now, we want to direct the reader to the work of Andrew Hall, who has published much material in video format on the Thunderbolts Youtube channel. He draws the conclusion that the Earth has been exposed in the past to severe electric and plasma events and argues that this shaped the geology of the Earth on a massive scale.

15.4.1 The dolomite problem

A long-standing "problem" in geology is the formation of dolomite. The creation of $CaMg(CO_3)_2$ is explained in general terms:

Dolomitization is a geological process by which the carbonate mineral dolomite is formed when magnesium ions replace calcium ions in another carbonate mineral, calcite. It is common for this mineral alteration into dolomite to take place due to evaporation of water in the sabkha area [coastal, supratidal mudflat or sandflat in which evaporite-saline minerals accumulate]. Dolomitization involves substantial amount of recrystallization. This process is described by the stoichiometric equation:

$$2\ CaCO_3\ (calcite) + Mg^{2+} \leftrightarrow CaMg\ (CO_3)_2\ (dolomite) + Ca^{2+}.$$

[Wikipedia 2021/Dolomitization]

The consensus is that the calcium is replaced by magnesium with water as the transport medium. How this would result in dolomite being "a double carbonate, having an alternating structural arrangement of calcium and magnesium ions" is not so easy to explain. The assumption that almost exclusively magnesium would be transported into the layers of rock and carry out the calcium does not really make sense. Both are soluble and even difference in solubility will not explain the layered feature of alternating calcium and magnesium nor the fact that only calcium would be transported out of the rock.

Another observation for dolomite is that "vast deposits of dolomite are present in the geological record, but the mineral is relatively rare in modern environments." This also goes against the idea that this is a common mineral that should still be produced today. In other words, there should be a simple observation possible where these processes are still occurring, yet this does not seem to be the case anymore.

What if we take a different position in our current accepted notion of the creation of the elements and open our mind to a new explanation with the knowledge that we have accumulated with SAM? Consider that the fusion with oxygen in biology is an often predicted reaction (Section 15.3) and looking at geology and mineralogy again, and in particular the case of dolomite, we can come to a very different conclusion:

The original $Mg(CO_3)$ or magnesite contained initially only the magnesium as a metal. When we use the reaction of Mg-24 + O-16 → Ca-40 we can see that the formation of dolomite may come from fusion "in situ." There are many more possible similar reactions, and we would have a very different explanation for the formation of many layers in geology.

15.4.2 Veins in geological features

A walk through a mountainous area will often show quartz veins running through the base rock. It has been argued that the first metals humans used were gold, silver, and copper, because they can be found in their natural metallic state in these veins. The weathering of the veins and bedrock is the reason we can find them in creeks and rivers from small flakes to actual nuggets of gold.

When posing the question "why are there so many (heavy) metals in these veins?" we hear the common response that they were created by hot pressurized water depositing certain metals at certain places. Such an explanation offers some problems though.

How can veins that are overly abundant and found all over the planet contain so many different types of metals in a concentrated form? Water transport and the chemical properties of these metals cannot explain this process properly. We would at least expect a much greater variation in the number of metals in such a system. In reality however we see that most of the ores found in these veins tend to be rather pure in composition, or a mixture of several elements that are often found together, such as silver, gold, and copper. Water, being able to tap into a supply of gold, dissolve lots of it, and transport it through the veins, depositing it at certain depths, is a little hard to accept. Especially the concentrations and absolute abundance of the metals is an enigmatic riddle, since the sheer amount of deposited metals would mean that an enormous amount of water must have passed through these veins while depositing the metal from the solution. There are no deposits of large amounts of gold specifically, or lead, or any other element that we know of anywhere else. The deepest holes that humans have drilled so far seem to produce predominantly silicates like mafic minerals (rich in iron and magnesium) and lots of water. In fact, the highest concentration of these metals is found in veins and not in such deep drill holes. We must conclude that the heavier elements are predominantly found on the outermost part of the planet, as far as we can observe.

Can transmutation explain the high concentration of specific metals in veins? The silver mines that were operational in the Ore Mountains (Erzgebirge) in Germany centuries ago were always following veins of silver ore deeper into the Earth. At some depth the silver ore stopped and the vein itself changed into a different type of dark ore that had no known use at that time. This ore was commonly referred to as *Pechblende* (pitchblende), which roughly means "brown/black (deceiving) ore." It was considered a "bad luck ore" since it always meant that the silver vein was depleted.

Centuries later, during the Cold War era, many of the known depleted silver mines suddenly became of interest again. The pitchblende, or rather uraninite as we know it now, is made up of the mineral U_3O_8. The ore consists of a high concentration of uranium and these mines were reopened and stripped of this ore centuries later. The Soviet Union used the mines in the Erzgebirge and the United States used the silver mines in the southwestern part of the country that had similar depleted silver mines in order to mine pitchblende.

Water, regardless of circumstances and conditions, is not known to be able to take certain metals and deposit them in an arbitrary place in a consistent manner. There must be another explanation. The idea that these metals can be concentrated with the help of water, pressure, heat, and the like cannot be maintained. Veins are the most important source for many of the metals we use in today's technological world. They contain the highest concentration of metals in (relatively) pure composition, are ubiquitous, and show certain features, like a silver vein turning into a uranium vein at a certain depth.

The idea that elements can be created in situ here on Earth sounds far-fetched or impossible to many. It is proposed that these veins are electric discharge channels, created under primordial conditions or during the catastrophes the Earth suffered repeatedly in the past. Such discharges would almost instantly ionize and even vaporize much of the material in the layers of the crust—as a result we see these veins in a lot of places. Breccia pipes seem to be caused by more explosive, larger events, creating pipes that are reminiscent of volcanic pipes, with large and small chunks of original material surrounded by metamorphosed material. We will look at breccia pipes in more detail in Section 15.4.3.

Each layer the discharge channel or vein pierces through has a different chemical composition and shows differences in other features as well. Some have lots of water, some are nearly dry, some are under more pressure, some have light elements, and some have heavy elements such as iron or barium. We suggest all these circumstances are of importance. Each layer would therefore—after an electrical event—have its own outcome in terms of in situ transmutations. Layer A would yield silver and layer B, underneath layer A, would yield uranium, for example.

This is a simplified view of reality though. We know almost nothing about this topic yet, since it is considered "inconceivable" by current geologists, chemists, and physicists. Anyone daring to think about such a topic, let alone get it published, is quickly dealt with by peer pressure objecting to such heretical ideas.

Currently, scientists are worried that groundwater in breccia pipe systems will pollute surrounding areas because the water in these systems is highly enriched in uranium. Uranium does dissolve. What it does not readily do is deposit again. The mechanism to explain such deposits is—as stated before—based on non-sensical arguments in consensus science. Water does not deposit uranium in any place unless it vaporizes leaving the uranium behind. However, in that case we would see all the other minerals in the water mixed in with the uranium ore—that is not the case.

Uranium ore, such as pitchblende, is always reported to have a high content of rare-earth elements and thorium, lead, and helium-4 as a result of decay processes. Apparently, natural decay of the radioactive elements yields rare-earth metals such as cerium and lanthanum. This is no surprise—we have known since 1940 that the fission of uranium can happen spontaneously in nature.

Why is there about 1%–2% silver in galena (lead glance), a PbS ore? Why is it silver and not some other metal? Why is the PbS surrounded by other quartz such as baryte and pyrite? Are these other elements simply condensing/crystallizing on this ore after the creation event of the PbS? It seems that the heaviest elements are always in the "center" of such things [Wikipedia 2021/Galena]. There has to be a logical and simple explanation to create those kinds of minerals. Even if it is chemical sorting, the current story does not add up. Most of the ore body is zinc-, lead-, silver-, gold-, and barium-rich. One example where galena and therefore silver is found is the Admiralty mining district in Alaska.

Also found in the same district is the Funter Bay underground lode mine. This mine produces about 500,000 tons of copper/nickel/cobalt ore, without gold, from a mesozonic gabbronorite pipe [Wikipedia 2021/Admiralty_mining_district]. Looking at the surrounding rocks/material we observe mostly elements like magnesium, iron, aluminum, meaning generally lighter metals—and silicon-oxide. The heavier metals such as gold are concentrated in the material that is considered an intrusive magma type coming from lower regions. Figure 15.1 shows a cross-section of what the geology of many ore mines looks like, although there is great variation in details. There is speculation that these structures are created as a result of electric forces from below plasmafying the material upward and transmuting elements in situ in the process. How and why this happens exactly remains at this stage a research topic.

The transmutations appear to keep fusion elements together until all energy is used up—in other words, when the fusion became less exothermic or even endothermic. We end up with elements clearly above the generally postulated iron fusion boundary, which is based on the shape of the curve of the average binding energy (Fig. 7.5).

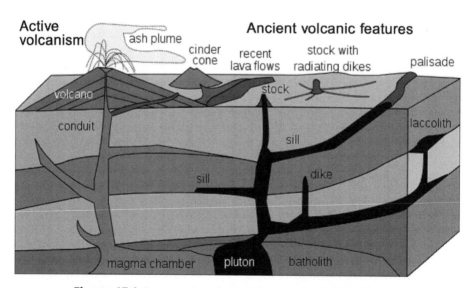

Figure 15.1 Cross-section of an intrusion creating dikes and pipes.
[https://en.wikibooks.org/wiki/HKDSE_Geography/M1/Intrusive_Vulcanicity]

Are the lighter elements forced into fusing to form heavier elements due to a massive energy surplus in an environment where there is enormous pressure such as a discharge in the deeper parts of the Earth? Are these lightning paths through the rocks creating veins? Metals such as (freshly created) gold, silver, and copper can, due to their nature, be deposited as a plating after the event in these discharge paths. They do not readily oxidize. If all lighter elements are used (and even the remaining overabundant oxygen is fused into sulfur), efficient fusion is no longer possible because there is a lack of fuel. When a mixture of titanium and oxygen has used all the oxygen and transmuted the titanium into nickel-62, we see an end-state, a nickel deposit. If the nickel still has a lot of oxygen available it would continue to fuse into 78 protons—selenium, for example. Since oxygen is usually overly abundant in the Earth's crust, it does not easily run out. So, we can see that under such circumstances the metals keep growing rapidly by fusing lighter elements such as carbon and oxygen onto it. This would explain why the creation of uranium or lead, which have a lot of "stress energy" or potential nuclear energy available for use, could be created in such abundance on the outer fringes of the planet. The electrical discharge would provide the energy to create heavier elements through fusion.

15.4.3 Breccia pipes

Breccia pipes are assumed to be ancient volcanic eruptions. However, their creation seems more likely the result of an electrical discharge of very large magnitude (as described above) rather than the result of an explosive eruption of material (breccia) based on volcanic activity.

In the top left of Figure 15.2 we find the label *petrified forest*. Around a breccia pipe we find fossilization and instant petrification at the edges, instant vaporization and plasmafication, and transmutation in the center, based on and related to layers of material. One other interesting detail in the illustration is that we see a uranium "plug" at a certain depth around a specific geological layer, namely the "Hermit shale" layer. This layer is rich in iron and many other elements due to shales that originate from mud and clay. The layers above and below the Hermit layer are more depleted of these "heavy" elements. There seems to be a direct correlation here that is currently not addressed or understood. We are of the opinion that this is a piece of evidence pointing in the direction of "in situ transmutations." Many of the more conventional physical creation processes are sure to have taken place during these events. However, we point out that in addition transmutations of the elements may have taken place.

This conclusion may seem "far-fetched" at first. However, this would explain not only the high abundance of heavy elements in the Earth's crust, and in particular within all vein-like structures, but also why they would be so prevalent on the very outer layer of the Earth? This in situ transmutation, if proven to be true, would change the entire collective paradigm of the creation of the elements and as a result we would need to take a very good look at our current models and accept that there may be other explanations to some basic questions. It also invalidates all those dating methods (e.g., radiocarbon dating based on C-14 or the potassium–argon method) based on initial creation or constant production abundance distribution and decay schemes because the base assumptions of these methods then are simply not true.

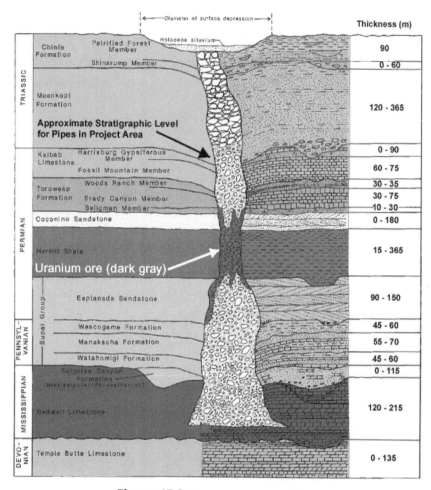

Figure 15.2 Breccia pipe cross-section.

[http://www.northern-arizona-uranium-project.com/breccia_pipe_anatomy, also
https://geotripper.blogspot.com/2012/01/uranium-mining-in-grand-canyon-perish.html]

These circumstances are most likely not taking place here on Earth in our era, although that may not be true for deeper regions below our feet.

15.5 TRANSMUTATION AND THE ABUNDANCE OF ELEMENTS

It is interesting to compare the abundance of the elements in the crust of the Earth with:

- the abundance values of "not crust," such as "oceanic floor," mantle, Moon, planets, Universe . . .; and
- the proposed or observed fusion reactions by Kervran.

Table 15.3 Example element abundance comparison.

Element	Abundance		
	Universe [PTE 2021/Properties/A/ UniverseAbundance.wt.html]	**Mantle** [Wikipedia 2021/Abundance_of_ the_chemical_elements#Mantle]	**Crust** [PTE 2021/Properties/A/ CrustAbundance.an.html]
Potassium	0.0003%	0.30%	1.5%
Silicon	0.065%	21%	27%
Magnesium	0.06%	22%	2.9%
Calcium	0.007%	2.3%	5%

We already provided some numbers in Section 10.1. A new question arises: Does life contribute to the abundance of certain elements in Earth's crust? This is a research topic. Do transmutation reactions inside the Earth (still) play a role in volcanic and black smoker activity and other geological processes like earthquakes?

Just by looking at the abundances in Table 15.3, it is clear there are great differences in the numbers. Magnesium has 24 protons and is quite light. Magnesium-24 plus oxygen-16 yields, as proposed, calcium-40 and adding oxygen-16 to carbon-12 yields silicon-28 (Chapter 12). Silicon and calcium seem to increase enormously in the continental crust in particular, while magnesium, the base element in these reactions, seems to be depleted in the crust in relation to what we think is in the mantle. It seems that these proposed reactions are not so strange as they may appear at first glance; the abundance of the elements in different mediums confirms this.

15.6 TRANSMUTATION IN EXPERIMENTS

We will now take a closer look at some examples of transmutation in laboratory experiments. We already mentioned that there is an interpretation of the LENR term much narrower than we have used until now. The term is used to describe a subset of experiments that are done in the laboratory that show transmutation (although even members of the LENR community deny this), excess heat, and produce no hazardous radiation. The last part about "no hazardous radiation" is a little bit questionable from our point of view. While there can be nuclear reactions causing transmutations that do not produce hazardous radiation (proton or deuteron capture at the right place) and many experiments might fall into that category, this is not the norm. We recommend being very cautious.

Before we look at such experiments we first need to look at a technique that is often utilized.

15.6.1 Electrolysis

During *chemical* electrolysis a direct electric current (DC) is passing through an electrolyte to produce chemical reactions at the electrodes. The main process is the interchange of ions by the removal or addition of electrons. Positively charged ions move toward the negative cathode, negatively charged ions move toward the positive anode. Electrons are removed at the cathode and introduced to the anode. When neutral atoms on the surface of an electrode gain or lose electrons they become ions and may dissolve in the electrolyte and react with other ions. The minimum voltage difference between anode and cathode of an electrolytic cell that is needed for electrolysis to occur is called the decomposition potential. The production of chemical reactions is proportional to the applied current above the decomposition potential.

But is that all? Ionized atoms are also becoming unstable [Filippov et al. 2006]. Why is that? We can think of the following reason: once some or all outer electrons are gone, there is pressure on the quasi-inner electrons to replace them. This in turn exerts a force on the endings or branches they connect to. It is conceivable that given that enough electromagnetic forces are involved and/or enough time this pull causes the nucleus to decay, fission, or even disintegrate. This decay/fission/disintegration could yield megaelectron volts, with only electron volts or kiloelectron volts put in.

15.6.2 Fleischmann and Pons—"Cold Fusion"/LENR experiments

A press announcement on 23 March 1989 in Salt Lake City, Utah, caused a worldwide sensation when two electrochemists from the University of Utah claimed they had observed nuclear reactions at room temperature. During the ensuing controversy, the term "cold fusion" became indelibly associated with it, even though the two scientists, Fleischmann and Pons, only claimed that their observations must have been caused by a nuclear reaction, since they couldn't imagine it to be caused by any known chemical process. Since then the term "cold fusion" has been replaced by various other names, but most used is Low Energy Nuclear Reactions (LENR). The field is still scientifically controversial, despite numerous experiments and hundreds of papers clearly signifying that the effects are real.

Four years before the mentioned press announcement Fleischmann and Pons had experienced a so-called "meltdown" in their university laboratory. They had been loading deuterium into a 1-cm cube of palladium by electrolysis. After months of measurements trying to characterize the behavior of deuterium in the palladium electrode, it exploded, vaporizing most of the palladium and burning or blowing a 1-ft-diameter hole in the table, as well as a 4-inch-deep pit in the concrete floor. They knew that the accident couldn't be a chemical one, and so they became convinced that the effect must be of nuclear origin, giving them the confidence to stand by their findings under the blizzard of attacks soon to be unloaded upon them.

A detailed account of the incident is given in Beaudette [2000, 33–43]. They redesigned the experiment changing the shape and size of the cathode minimizing the amount of palladium available for meltdown. A schematic of the Fleischmann and

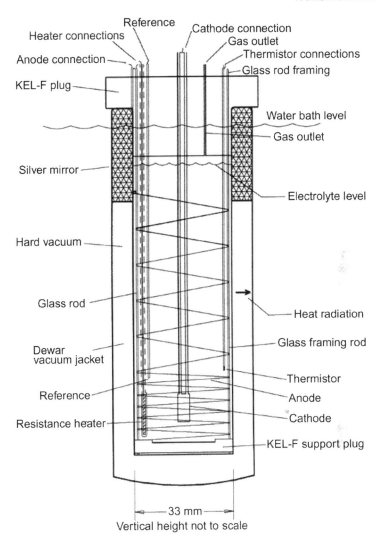

Figure 15.3 Fleischmann and Pons cell schematic layout.
[Beaudette 2000, 37]

Pons electrolytic cell is shown in Figure 15.3, reproduced from Beaudette [2000, 37]. Here is a summary of the cell's operation as documented by Beaudette:

The electrolyte was heavy water with lithium metal dissolved in it. . . . The thermistor (for temperature measurement) and the calibration heater (for inserting electrical heat pulses) are used to calibrate and monitor the rate of radiant heat transmission by the cell to the surrounding water bath. When electric current flows through the cell, the water molecules break up into two gasses. Oxygen is produced at the anode (+) and deuterium at the cathode (−). These gasses bubble up to the surface and leave the flask. The energy that was put into the electrochemical reaction that separated the water molecules into the two gasses is carried away from the cell with the gasses. Electrolyte solution is added each day to make up for what was bubbled away during the previous twenty-four hours.

. . .

Once the cell current is turned on, it ordinarily operates continuously day and night for weeks until the end of its life . . . The most critical part of the cell was the surface of the cathode electrode. The excess energy that was claimed would be generated in or near its surface. Many researchers have experienced a buildup of unknown material on the surface of the cathode that the electricity could not get through, bringing the experiment to an end. These problems have always plagued electrolytic experimentation. [Beaudette 2000, 36–38]

Thus, Fleischmann and Pons' primary claim was the detection of excess heat. They didn't find the nuclear products that were expected in hydrogen–helium fusion. Although some amount of tritium was detected, no neutrons or gamma rays were observed. Since the excess heat effect was rather subtle, virtually no one could replicate Fleischmann and Pons' results in hastily put together experiments in the weeks and months following 23 March 1989, also because there was no accurate description of the Fleischmann and Pons apparatus and the experimental procedures employed. The nuclear physics critics at the time basically dismissed the heat effect while looking for the nuclear products that they were sure had to be there. This failure to find confirmation of the anomalous heat effect (AHE) at places like Caltech, MIT, Brookhaven, and Harwell (UK) was used by mainstream science to dismiss the effect as either a deliberate hoax or the result of incompetence of the experimenters.

It was only years later that Pons showed anomalous deposits on the electrodes of the Fleischmann and Pons reactors. Like in many other similar experiments in the next three decades, the conclusion was that transmutation of the electrode had occurred during the excess heat episodes, although no one drew that conclusion in the early years of "cold fusion."

Not surprisingly, within a few months of the Fleischmann and Pons press announcement, the science community turned en masse against Fleischmann and Pons, characterizing cold fusion as pathological science, essentially destroying their careers.

We have already established that electrolysis running long enough could cause nuclear reactions (Section 15.6.1). In combination with other factors we could see transmutations and released heat. This process can be greatly expedited and controlled by deuteron capture if the given material is receptive to it. Unfortunately, the isotopic composition of the palladium cathode is not given.

15.6.3 Mizuno and Miley experiments

In 1997 a book was published in Japan with the title *Nuclear Transmutation: The Reality of Cold Fusion* [Mizuno 1998], translated in 1998 by Jed Rothwell. In this

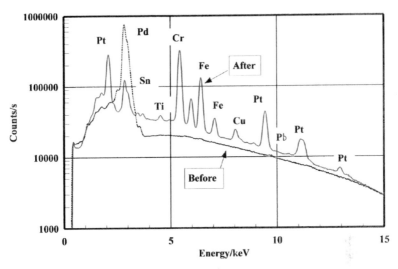

Figure 15.4 An example of an EDX [energy dispersive X-ray spectroscopy] element analysis of a palladium rod after it produced excess heat during electrolysis. Results from before and after electrolysis are shown. In the palladium rod after electrolysis, all sorts of elements are observed.
[Mizuno 1998, 72]

book Mizuno gives a detailed account of his LENR experiment findings. In contrast to the Fleischmann and Pons' electrolytic cell, Mizuno's experiment operated at a pressure of around 10 atmospheres and at much higher power densities. In one incident in March 1991 an "anomalous heat burst" occurred, described at length in the book. Mizuno had to go to great pains to remove all the stored energy by dumping the experiment in a bucket of water that kept evaporating for over a week. In the end the accumulated excess heat amounted to as much as 100 MJ.

Apart from the production of massive amounts of excess heat, Mizuno observed the appearance of new species of elements that were not there before the electrolysis. A typical result is shown in Figure 15.4. The presence of large amounts of chromium, iron, manganese, and copper in the post-analysis of the palladium cathode cannot be dismissed as coming from contamination, as many critics have attempted to do.

A third major issue shows up in the exposed cathode in the form of changes in the ratios of the six stable isotopes of palladium from the naturally occurring abundances. This is shown in the lower part of Figure 15.5. Norman Cook has performed an analysis of these "bizarre patterns of both increases and decreases" through a straightforward simulation of the observed effects [Cook 2010, chap. 9]. He concludes that "the pattern is far from bizarre—and indeed the simulation shows the participation of *all* palladium isotopes to about the same degree in the 'cold fusion' reaction . . . It therefore appears that there are no 'special isotopes' of palladium—comparable to the special role of U-235 in the fission of uranium." Thus, monoisotopic experiments as proposed early on by Mizuno should in his opinion be unnecessary.

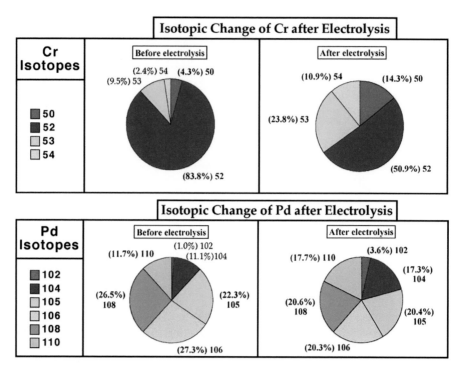

Figure 15.5 Changes before and after electrolysis with Cr and Pd. Isotopic abundances in percentage are shown in pie charts. For Cr, natural and post-electrolysis abundances are shown (because there was no significant Cr in the sample before electrolysis). For palladium, abundances before and after are shown.
[Mizuno 1998, 72e]

Figures 15.6 and 15.7 show the mass spectrum of experiments performed by Mizuno in 1992. Many elements that are absent prior to electrolysis appear in the analysis of the cathode after the experiment.

Unfortunately, the percentages given before and after for chromium and palladium are not very helpful, absolute numbers would be better. For the same reason Norman Cook's simulation (see p. 209) is not expedient, and we disagree with his assessment, based on the structure of the nucleus in SAM. Monoisotopic experiments are very important as some isotopes of an element are more receptive to proton capture than others.

15.6.4 Urutskoev—exploding titanium foil

The Urutskoev experiments were done in 2002 by Urutskoev at the Kurchatov Institute in Moscow [Urutskoev et al. 2002]. In the example shown in Figure 15.8 a thin titanium foil was exploded underwater using electrical discharge. It shows that new elements were created, including aluminum, silicon, calcium, titanium, chromium, iron, nickel, copper, and zinc. Also shown is the change in isotope ratios before

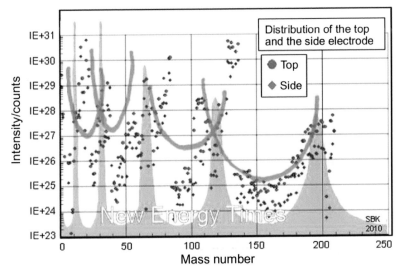

Mass spectrum of SIMS measurement for Pd sample surface, which
generated excess heat, after electrolysis in 1991. Confirmatory tests also
with EDX, AES, EPMA, ICP-MASS. Two areas of surface were analyzed and
show similar spectra [1]. Upper 5-peak peak green curve is an overlay from
Miley experiment, drawn by Miley for his data in 1996 [2]. Lower curve,
shaded in purple, drawn by Larsen in 2000 with no fitting, based on
Widom-Larsen ultra-low momentum neutron absorption model. [3]

1. Mizuno, Tadahiko, "Isotopic Changes of Elements Caused by Various Conditions of Electrolysis,"
American Chemical Society, March 2009
2. Miley, G.H, and Patterson, James, "Nuclear Transmutations in Thin-Film Nickel Coatings Undergoing
Electrolysis," Journal of New Energy, Vol. 1(3), pg. 5, (1996)
3. Larsen, Lewis, Feb. 7, 2009 slides

Figure 15.6 Mass spectrum of Mizuno data compared with Miley data and Widom–Larsen
Theoretical Model.
[Krivit 2012, 77]

and after the experiment. The titanium-48 [Ti-48] content went from 73.8% to 65%,
whereas for the other titanium isotopes the ratio went up.

The Urutskoev experiments seem to represent a forced disintegration of material
through electric discharge. Pieces of atoms are broken off and re-fuse to bigger pieces
in a more or less chaotic process.

15.6.5 The Mitsubishi Iwamura LENR gas permeation experiments

The LENR gas permeation experiments were performed by Iwamura of Mitsubishi
Heavy Industries in 2002. The operation is briefly described in the *New Energy Times*
Special Report cited above regarding the Mizuno data:

*The essential part of the experiment is a multilayered substrate of palladium
and calcium oxide [Figure 15.9]. On the surface of the substrate, the Mitsubishi*

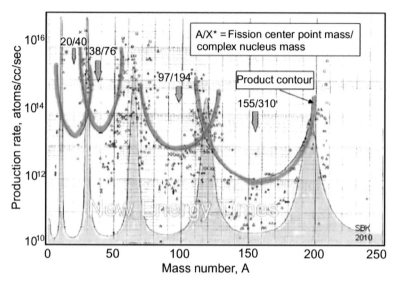

Upper 5-peak peak curve drawn by Miley in 1996 based on 6
experimental runs of transmutation yields from Ni-H LENR systems [1].
Lower curve, shaded in yellow, drawn by Larsen in 2000 with no fitting,
based on Widom-Larsen ultra-low momentum neutron absorption
model. [2]

1. Miley, G.H, and Patterson, James, "Nuclear Transmutations in Thin-Film Nickel Coatings
Undergoing Electrolysis," Journal of New Energy, Vol. 1(3), pg. 5, (1996)
2. Larsen, Lewis, Feb. 7, 2009 slides

Figure 15.7 Mass spectrum of Miley Mizuno data compared with Miley data and Widom–
Larsen Theoretical Model.
[Krivit 2012, 76]

*researchers place atoms from a given element. Deuterium gas is passed through
the substrate. As the experiment progresses, the given element is seen to decrease
in quantity, and an element which was not present before the experiment slowly
appears on the surface and increases in quantity.*

[. . .]

*They have repeated this type of observation many times, with several pairs of
elements:*

133Cs	→	*141Pr* (addition of 4 deuterons)
88Sr	→	*96Mo* (addition of 4 deuterons)
137Ba	→	*149Sm* (addition of 6 deuterons)
44Ca	→	*48Ti* (addition of 2 deuterons)

[. . .]

*The gradual increase of one element and the temporally correlated gradual
decrease of another element are consistent features of their experiment. As shown
below [Figure 15.10], their XPS data create a similar pattern among three sets
of experimental runs. Again, the gradual, temporal signature of growth and
reduction is crystal clear. [Krivit 2012, 35–39]*

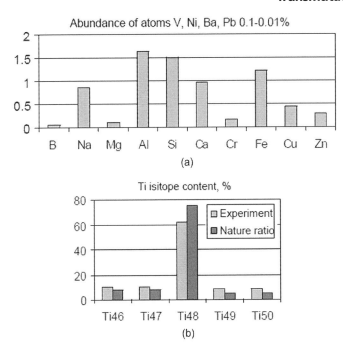

Figure 15.8 Results of mass spectrometer analysis of products in test 226 (Ti load).
(a) Percent composition of atoms of "alien" elements in the sample. The fraction of Ti atoms
in the experiment products is 92%. (b) The ratio of Ti isotopes before and after the experiment.
[Urutskoev et al. 2002, 711]

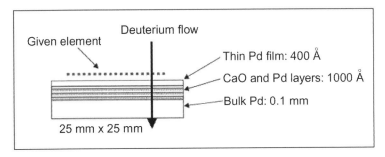

Figure 15.9 (*Left*) Conceptual view of multilayered substrate. (*Right*) Detailed view of
multilayer fabrication—five layers each of Pd followed by CaO, finished with a final layer of Pd.
(Adapted from [Krivit 2012, 35])

Interestingly, the step from cesium-133 to praseodymium-141 involves a five-deuteron
difference. Barium-137 to samarium-149 involves a seven-deuteron difference.
Calcium-44 to titanium-48 involves a three-deuteron difference. In each of these
cases, additional resettling happened to create a missing deuteron from PEPs including $\beta-$ decay.

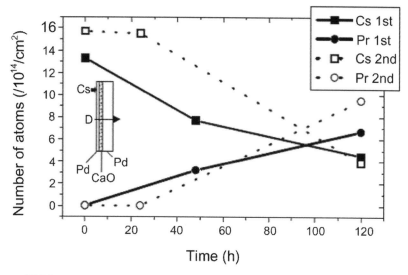

Figure 15.10 Two experimental runs showing temporally correlated gradual decrease of cesium and increase of praseodymium.
[Krivit 2012, 36]

Unfortunately, there is no information available on what happened to the palladium substrate during those experiments. We would expect to see some of the palladium being transmuted to silver.

15.6.6 The SAFIRE project—a current case study

SAFIRE stands for Stellar Atmospheric Function in Regulation Experiment. Its goal was to test the Electric Sun model as part of the Electric Universe theory.

At the beginning of 2012 at the *Electric Universe Conference* in Las Vegas, Montgomery Childs first presented his idea to test the Electric Sun model. The idea was to create a Solella Experiment comparable with Kristian Birkeland's Terrella experiments at the beginning of the 20th century. To keep the operation efficient, a Design of Experiments methodology was to be used. With this methodology the number of experiments can be kept small once the primary factors for the test have been identified.

The project was given the go-ahead in 2013 and a team was created. In January 2014 the first bell jar experiments (phase I) were done. The main goal of these experiments was to collect data from the experiments to be used in the design and engineering of a larger reactor. The tests revealed self-organizing plasma double-layers. It was determined that the shape of the cathode does not play a significant role in the formation of plasma double-layers. Already in this first experiment, optical spectroscopy indicated new elements on the anode. Also, anode tufts were observed, as well as the medium energy trapping of electrons, ions, and molecules. After these encouraging results, phase

II began—the design and engineering of the large reactor. Design and engineering of phase II was completed in 2015. Phase II with the accompanying control room/laboratory was constructed and tested in 2016, with preliminary results shown within three months. The reactor chamber has an anode made from various heavy alloys in the center of the chamber through which gases can be pushed (e.g., hydrogen). Additionally, the chamber can be evacuated or filled with gases. The cathodes are located in the outer region of the chamber. The capabilities of the phase II reactor are listed as follows [Aureon Energy 2021/Reactor]:

- stable self-organizing plasma double-layer shells
- stable plasma tufts analogous to stellar plasma tufting in the photosphere
- energy densities analogous to the Sun
- electromagnetic confinement of matter
- steep voltage drop just off the surface of the anode
- acceleration of ions from the SAFIRE core
- spectral line broadening showing higher energies in plasma corona
- uniform thermal radiation emission—low thermal buoyancy
- creation of concurrent collisional and non-collisional plasma
- eruptive discharges analogous to Solar CMEs
- chemistry as a catalyst to double-layer formation
- slowing the speed of UV light by 5,000 times
- analogous transformer/capacitor behavior.

In 2017, extensive studies were done on plasma double-layer formation. It was discovered that certain gas species work as catalysts in the formation and stabilization of plasma double-layers. First indications of transmutation and high energy output emerged. A SEM/EDAX (Scanning Electron Microscope/Energy Dispersive Analysis X-ray) analysis showed new elements—transmutation was suspected. Observations were made of energy discharges analogous to the Sun's. A high voltage drop across 20 microns was confirmed close to the anode. 2017 also brought a surprise—melting tungsten Langmuir probes with changes to the structure of the material as well as indications that the cooling system of phase II might be insufficient. Further research was conducted into plasma double-layers and plasma tufting. In 2018 medium-energy plasma discharge experiments were conducted, also analysis and modeling using various gas compositions were done. Dark-mode plasma structures were discovered. In 2019 ongoing experiments tested the thermal limits and boundaries of the chamber. New gas compositions and anode alloys were introduced. Additional experiments and tests were done to quantify the dark-mode plasma. It was confirmed that certain catalyzing elements will transmute elements in medium-energy plasma regimes. It was also confirmed that the cooling installed on the phase II chamber was indeed insufficient. SEM/EDAX third-party lab tests confirmed elemental transmutation in the chamber. Phase II ended in 2019. No disparities with the Electric Sun model were found, SAFIRE was declared a success. The focus for the future transfers to commercialization (heat production, nuclear waste remediation, etc.) of the reactor [Aureon Energy 2021/History].

The important question regarding the experiment is: What really happened starting in 2017 when the first Langmuir probe (with a tungsten tip) vaporized in an instant and the first signs emerged that the thermal boundaries of the chambers might be put to the test? The "misbehaving of the chamber" started already back then—not in early 2019. The results of this last round of experiments in 2019 with the phase II chamber were presented at the *Electric Universe UK 2019 Conference* in Bath (UK).

We will now have a more detailed look at the results presented at the conference. We assume that atomic hydrogen was pushed through the anode. During a test run with various regimes the anode suddenly melted and the thermal limits of the chamber were reached with just 7% of the maximum input energy. The experiment needed to be shut down. The coloring of the anode clearly showed that something happened there. SEM/EDAX analysis revealed the following elements to be present on the anode after the experiment. The second column shows the typical number of protons of the new elements found:

- Carbon   (12–13)
- Oxygen   (16–18)
- Sodium   (23)
- Magnesium   (24)
- [Aluminum]   (27)
- [Silicon]   (28)
- Phosphorus   (31)
- Sulfur   (32)
- Chlorine   (35)
- Potassium   (40)
- Calcium   (41)
- Titanium   (46–50)
- Zinc   (64)
- Tin   (114)
- Barium   (132)
- Lanthanum (predominantly)   (139)
- Cerium (predominantly)   (140)

Although the elements in square brackets were mentioned, they were discarded as potential contamination. In the atmosphere of the chamber the following elements were found by optical spectroscopy:

- Lithium   (7)
- Sodium   (23)
- Manganese   (55)

An additional three elements were found on the anode but were not identified in the presentation. Also, much of the optical spectroscopy still needs to be analyzed. There were probably many more elements in the chamber atmosphere that have not been identified yet.

The important questions about this experiment are:

1 Which materials were used in the anode?
2 What was the "catalyst"?

Before we continue, we must ask another question: Where have we seen barium, lanthanum, and cerium as a product of a nuclear reaction before? Read Section 13.1 again for the answer. Coincidence? We do not think so. What did they do in SAFIRE?

The tungsten Langmuir probe incident of 2017 is from our point of view the key to answering the first question. While sitting around a fire at Pepperhill Barn, waiting for the next part of the 2019 EU conference to begin, an idea was born: What if the experiments with *tungsten* continued in 2019? It would be the logical choice after what happened to the probes—at least for us. And so a presentation was given the next day at Pepperhill Barn with this assumption in mind. These events and our conclusions back then are documented in a conference report [Otte 2019].

We will now reiterate and refresh our conclusions from back then, assuming that the anode was at least laced with, but most likely was made out of tungsten. Commercially available tungsten can be purchased in the form of alloys usually with at least 90% tungsten, the rest is made up of nickel, copper, and iron in various combinations. Tungsten purity can go up to 97%.

Tungsten (tungsten-182) looks like Figure 15.11 in SAM. Tungsten-182 is very receptive to proton capture, all six branches would accept protons easily on the five-endings. In the experiment, hydrogen-1 is pushed through the anode. Once an electric field is established, the hydrogen would be stripped of its outer electron and we would see a stream of protons (with some kinetic energy) pushing through the metal lattice of the anode material(s). This fits the description we made for potentially successful proton capture (Section 11.11).

An incoming proton combines with one PEP to create a deuteron. Examining the SAM nuclear periodic table of elements in Appendix B, which also includes unknown, missing elements, we can see that one step up from tungsten, there is a missing element which is shown in Figure 15.12.

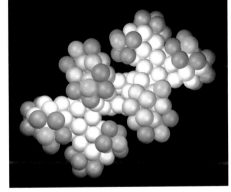

If the proton hits one of the five-endings that already show a lithium nuclet, the second lithium nuclet on one branch in this configuration is not stable. Then we would have a lithium nuclet opposing the other branch with a lithium nuclet such that the isotope disintegrates. The same thing happens when the five-ending on the carbon is hit. We end up with an unstable configuration.

Let us further assume that an electromagnetic environment (ionization) in combination with proton capture

Figure 15.11 Tungsten-182 (*top view*).

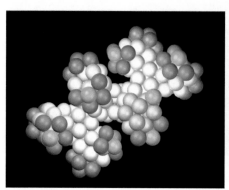

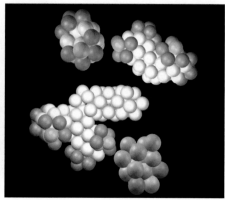

Figure 15.12 Missing element after tungsten with 183 protons.

Figure 15.13 Disintegrating isotope with 183 protons.

is a stressful situation for an element and is potentially capable of breaking up the structure. The easiest point of collapse for the structure are the anchor points of the branches, similar to what we saw during fission with heavier elements like uranium. There are also several options of branches fusing together during the process as a secondary reaction. This is an educated guess, supported by the quasi-inner electrons pulling branches together and by analyzing the isotopes found after the experiment.

Multiple branches can be removed (Fig. 15.13), one has for example 22 protons and one has 17 protons. If one of the outer branches is hit, the piece will have 23 protons. Overall 39 to 45 protons are broken off, which leaves 138 to 144 protons as the bigger part. So here we could have lanthanum-139 or cerium-140 and sodium-23, and the remaining 22-proton piece could easily capture another proton in this environment and transmute to sodium-23. There are also two possibilities to remove 55 protons as a branch. Here we have manganese-55 after rearrangement. We would therefore predict manganese to also be found on the anode. The electron affinity of manganese might be the reason for it to be present as well in the atmosphere of the chamber, maybe even on the cathode. Two 22-proton pieces can also fuse to calcium-44. Another piece has 17 protons, this could be oxygen-17. Carbon-12 as a broken-off branch piece is another possibility. There are many options to consider including how they could potentially fuse in a secondary step. For example, a branch could also come off while breaking one proton out of the backbone. If two were broken out of the backbone, we would see barium-138 as the bigger part. The lighter pieces then could potentially fuse, and we would see titanium-46. But what about the zinc detected? If the tungsten alloy also contained copper-63, which is receptive to proton capture, the result would be zinc-64. Or another fusion step with an 18-piece broken-off branch on top of titanium-46 is thinkable. Lanthanum, cerium, oxygen, and carbon are specifically mentioned in the latest SAFIRE presentations. Carbon is a logical outcome since it represents the highest binding energy of a nuclet, the densest packing, and the highest protons-per-volume ratio.

If it is possible to disintegrate atoms even further by even more fission steps, then carbon would be one of the most abundant outcomes, as it is a basic nuclet.

> In essence, we postulate that nuclear fission is the dominant reaction in the SAFIRE experiment of 2019. This does not exclude secondary fusion steps, but the main process is *nuclear fission of tungsten induced through proton capture in a highly charged electrical environment*.

This probably needs a few moments to sink in and the repercussions to become clear. All this is based on the assumption that the experiments with tungsten continued after the 2017 probe incident. However, we also have to consider that the experiments continued with random materials. But even then tungsten, with its high melting point, would be a good choice. We stand by our initial assumption: based on the model and the resulting observed transmuted elements, we conclude that the anode in the 2019 experiment presented in Bath was made from tungsten or at least was laced with it.

How much energy would be released by this proton capture/nuclear fission process according to SAM (Table 15.4)? As there is no literature information available on the unknown element with 183 protons, we chose a stress energy difference based on comparable isotopes existing in the range. Again, we look at this as a fusion energy balance sheet first.

Also, we would recommend installing real-time beta radiation detectors (Section 13.5). Tungsten has 18 quasi-inner electrons, cerium has 12. The biggest other pieces have a maximum of two. With every fissioning tungsten nucleus, four to six electrons are being released.

Could fusion be the primary action within SAFIRE? If fusion really happened by building smaller atoms on top of each other then we should see many more evenly

Table 15.4 Binding energy for cerium-140/2 × sodium-23 fusion.

Fusion reaction components	Total BE (MeV)	SAM BE (MeV)	Difference BE (MeV)
Cerium-140	1172.68	1261.58	−88.89
2 × Sodium-23	2 × 186.56	2 × 180.23	12.56
Cerium-140 + 2 × Sodium-23	1545.80	1622.04	−76.24
Tungsten-182	1459.33	1639.83	−180.5
Unknown-183	~1461.50	1646.50	~−185.00
Unknown-183 − (Cerium-140 + 2 × Sodium-23)	~ −84.30	24.46	~−108.76

In reverse, roughly 85 MeV per atom are immediately released in theory during this fission process. No wonder the anode melted and the chambers thermal limits were exceeded—assuming enough nuclei participated in the reaction.

distributed elements, especially if we think of hydrogen or deuteron fusion. If we consider larger atoms such as nitrogen, oxygen, and iron, then we should still see many elements reflecting that, which we do not. A single step for a transmutation (e.g., sodium to magnesium) is a logical, viable, known step, and it works pretty much all the way through the PTE. However, to create lanthanum or cerium from a single step fusion reaction would not be logical since then we would see again a much more evenly distributed element range in the results. The only logical reason for elements like calcium, titanium, lanthanum, cerium, and barium is fission. But this does not exclude secondary fusion steps with for example oxygen-17 with other fission results.

Deuteron fusion with tungsten is another process option instead of proton capture. However, the result is essentially the same. The deuteron latches on and the tungsten nucleus fissions.

The second question was about the "catalyst." What role does it play? We think it has nothing to do with the transmutation step itself, which is caused by proton capture. However, it does have something to do with creating the right electrical environment for the proton capture to be successful. This is in part based on the usage of the word "catalyst" on the *Aureon* website. Paul Anderson—another member of the SAFIRE team—mentioned in his presentations in previous years that some gases (e.g., oxygen) in the chamber basically kill the electrical environment inside SAFIRE while others seem to support it.

Our experiment description in Section 5.2 regarding electron affinity was based on this thought, and we stated that, say, helium, nitrogen or argon—based on their electron affinity—would be good candidate gases for facilitating the kind of electrical environment that is needed to support proton capture. We therefore submit helium, nitrogen or argon as possible "catalysts" inside SAFIRE when the assumed tungsten fission reaction happened. Oxygen, on the other hand, would most likely kill off the electrical environment, recombine with the protons from the anode, and just create water in the chamber, without transmuting anything.

There is also the claim that SAFIRE can do nuclear waste remediation. If we look at patents about nuclear waste remediation, we see at least one that operates in a way similar to SAFIRE. Both create a highly electrical environment and streams of protons pushing them through or over material—in this case radioactive waste. If a waste product is receptive to proton capture, the radioactive material might be transmuted to other, less radioactive or much faster decaying products or even stable elements. So, in theory this could work, but the practicability is at least questionable.

The three most relevant fissile isotopes are uranium-233, uranium-235, and plutonium-239. For fission reactors the fuel (typically based on uranium) is usually based on the metal oxide; the oxides are used rather than the metals themselves because the oxide melting point is much higher than that of the metal and because it cannot burn, being already in the oxidized state. In nuclear waste, 96% of the mass is the remaining uranium; most of the original uranium-238 and a little bit of uranium-235. Usually uranium-235 would be less than 0.8% of the mass along with 0.4% of uranium-236. Typical long-lived decay products are Sr-90, Sn-126, Cs-137, Tc-99, and

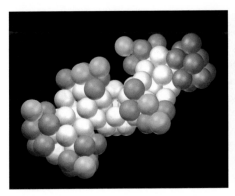

Figure 15.14 Technetium-99 (*top view*) is a common fission decay product.

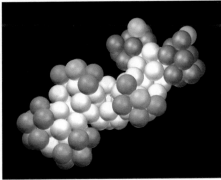

Figure 15.15 Ruthenium-100, proton capture product.

I-129. All of them and of course uranium itself are receptive to proton capture according to SAM. Let's have a detailed look at technetium-99, which usually β– decays to ruthenium-99 with a half-life of 2.111×10^5 years (Fig. 15.14).

Technetium-99 is receptive to proton capture with its boron ending and a lot of PEPs (Fig. 15.15). Some PEPs also change their location while a second boron ending is created.

This intermediate step rearranges to ruthenium-100 (Fig. 15.16).

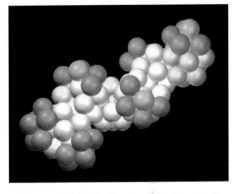

Figure 15.16 Ruthenium-100 (*top view*).

We conclude that at least for waste products that are receptive to proton capture the process can work. In the example above a long-lived β– decay waste product was reduced to a stable element.

In 2002 a patent was granted to the State of Oregon on behalf of the "Portland state university" called "Processing radioactive materials with hydrogen isotope nuclei" [Dash 2003]. This patent shows that uranium metal samples exposed to hydrogen isotopes, ionized through electrolysis, for hours, weeks, and months caused the uranium samples to have a hugely increased decay rate. The patent explains how the samples in relation to the control sample show more than significant increase in specific radiation signatures such as gamma rays. The samples also showed a clear decrease in the original uranium percentage. In effect, they calculated that the exposure changed the decay rate of uranium-238 from 4.5 billion years to 100 years.

15.6.7 Conclusion on experiments

Many experiments have been performed since 1989 in a multitude of experimental setups. Some provide no definitive results, yet a significant number show results that cannot be explained away. The fact remains that elements are consistently observed in these experiments, which were not there before the experiment, leading to the unavoidable conclusion that transmutations do take place, in relatively benign circumstances. Most of these elements pointing to transmutations are quickly dismissed as "contamination or accumulation of contamination." Yet when we take a better look at these experimental results, a clear picture emerges. Successful experiments have the following singular characteristic: An electrical environment with strong electric fields, a plasma, or showing at least some ionization.

There are several options to achieve this, electricity being the most common one, but laser light, sound, pressure and/or heat, or Ohmasa gas also seem to achieve this state. Common observations are:

- Transmutations observed (chemical analysis/spectroscopy).
- Energy surplus (exothermic).
- Accidents have happened in experiments with no explainable cause unless nuclear forces are involved (no chemical energy, or component in the experiment).
- Fusion as well as fission steps are observed.

From the above small sample of diverse experiments, all pointing to transmutations of one kind or another, it is hard to deny that observations of LENRs are real. That the field is still scientifically controversial despite hundreds of experiments three decades since the Fleischmann and Pons' press conference is largely due to the fact no acceptable theory has yet emerged.

We are aware that there are some interesting phenomena reported from LENR experiments that we have not addressed in this book such as plasmoids (EVOs—Exotic vacuum objects), erratic behavior, arc discharges, strange radiation, and electric field enhancement. However, the main focus of SAM is the nucleus, and we see enough evidence that transmutation and surplus energy released during the experiments can be explained through the change in binding energy of the atomic nuclei once the process happens. We consider a plasmafied environment a necessary component of the process to induce a reaction in the nucleus. Where the plasma is, transmutations are happening.

15.7 A SMALL PEEK INTO THE LENR COMMUNITY

After the news broke about the Fleischmann and Pons' experiment, many tried to replicate it. Eugene Mallove was at that time the chief science writer at MIT. When he found out that the results of the hastily performed experiments were presented after the data had been manipulated to show that there was no result at all, he quit his position and stated openly that scientific fraud was being committed. Mallove wrote a book called *Fire from Ice*, and he became perhaps the biggest proponent for what was then called "cold fusion." He gave many presentations about the topic pointing out that this is a real phenomenon, even if poorly understood. Too much evidence was pointing in this direction. Unfortunately, he was killed in a burglary in 2004 and this represented a huge loss for the field.

We have already mentioned a few names in the field of LENR in this chapter so far. The earlier mentioned researcher Nancy Bowen (see Section 14.1) belongs to the LENR community too. Another person to mention as an important part of the LENR community is David Nagel. He challenges the LENR advancements and the community with his questions [Nagel 2018].

Usually, there are a few conferences and/or workshops each year where the proponents meet and discuss progress. However, real progress in the field has been hampered for years now by the attempt to solve the LENR phenomenon within current physics.

15.8 SAM AND LENR

We are convinced SAM is the best model foundation for a LENR theory that is only partially developed.

It has become clear that a nucleus can be more readily changed than we once thought. We think LENR is a mixture of observed nuclear reactions that occur in many ways and through many mechanisms. A nucleus (in SAM) can under specific circumstances and when being hit at the right place, accept protons, electrons, PEPs, and bigger nuclei from elements such as deuterium, helium, carbon, and even oxygen, leading to a nuclear reaction that does not have to emit gamma rays, beta radiation, and/or PEPs. However, this is an exception. We cannot hope to explain LENR with the current models and theories. The results of applying SAM to LENR experiments are very encouraging in our view. So far, we have not found a single LENR experiment with enough data available that could not be explained based on SAM. Some of those experiments seem to be a combination of fission and fusion steps in quick succession when a nucleus resettles. SAM shows nuclear reaction energies from low yield to extremely high yield—based on the energy differences between the binding energy values before and after the transmutation step. The timeframes of the experiments depend on the setup—especially regarding the plasma environment—and can range from fractions of a second to several days/weeks/months.

> The energy surplus of LENR is all about transmutation and the resulting difference in binding energy. There is a lot of energy hidden in the nucleus and sometimes not much energy is needed to cause a nuclear reaction and, in the process, tap into that stored energy.

To proceed further with LENR, we need to execute some experiments with SAM in mind.

15.9 EXPERIMENTS FOR A BREAKTHROUGH IN LENR

In order to try to validate, or better falsify, SAM, LENR experiments can be conducted that will either comply with SAM expectations or refute them. If it is the latter, then we learn something new and the model will have to be adapted. We will assume a setup like SAFIRE with a "catalyst" that allows for proper electrical interaction.

Highly receptive to proton/deuteron capture:

- Nickel-62 $\xrightarrow{p}$ Copper-63
- Palladium-106 $\xrightarrow{p}$ Silver-107
- Palladium-105 $\xrightarrow{p}$ Silver-107
- Sodium-23 $\xrightarrow{p}$ Magnesium-24
- Potassium-39 $\xrightarrow{p}$ Calcium-40
- Aluminum-27 $\xrightarrow{p}$ Silicon-28
- Boron-11 $\xrightarrow{p}$ Carbon-12 (but may show α decay instead)

Less receptive to proton/deuteron capture:

- Nickel-58 $\xrightarrow{p}$ Cobalt-59 (after resettling)

Receptive, but not stable:

- Palladium 105 $\xrightarrow{p}$ Silver-106 $\xrightarrow{\beta+}$ Palladium-106
- Nickel-63 $\xrightarrow{p}$ Copper-64 $\xrightarrow{\beta+}$ Nickel-64
  $\xrightarrow[\beta-]{}$ Zinc-64

Not receptive to proton capture:

- Iron-54
- Carbon-12
- Palladium-102.

Various carbon and oxygen fusion options.

CHAPTER 16

A deeper look at LENR

16.1 INTRODUCTION

LENR can be viewed as a jigsaw puzzle with various missing pieces. There are several improbable observations in the field that require explanation. Even if an observation only occurs infrequently, it cannot be ignored because it might be a clue to one of these missing pieces.

In a 2018 review article, entitled "Expectations of LENR Theories," David Nagel provided a list of 10 questions and another list of 24 observations from LENR experiments that require theoretical explanation. Quoting from the paper:

> *The nearly three decades of theoretical work on LENR has resulted in remarkably few well-developed theories. None of them has yet been adequately tested and widely accepted . . . many (most??) of the theories would be eliminated by one or more of the listed observations. Unfortunately, many LENR theories fail at the conceptual level.* [Nagel 2018]

One important reason the matter has not been resolved was expressed by Hagelstein:

> *. . . any process must presumably be functioning in a manner not previously expected (lest it would have been found earlier).* [Hagelstein 1992]

We assert that a fundamental reason such a process has not been discovered is the incorrect application of quantum mechanical rules to the nucleus. This so-called Copenhagen interpretation of quantum mechanics does not allow fixed positions for the nucleons because that would be in violation of the Heisenberg Uncertainty Principle.

In contrast, the makeup of the SAM nucleus is ruled by the densest packing principle, resulting in each element or isotope having a unique geometrical structure, implicitly rejecting the Copenhagen interpretation. By adopting the view that the components of a stable nucleus have specific coordinates in 3D space, "all nucleon–nucleon interactions for any given number of protons and neutrons in the lattice can be calculated" [Cook & Dallacasa 2014, 75]. The *Structured Atom Model* is in our opinion the tool that enables discovery of the unknown process mentioned by Hagelstein.

16.2 SUMMARY OF THE SAM MODEL FOR LENR

Many experimentally obtained results in atomic/nuclear physics cannot be explained by any of the current models. This point is addressed in Chapters 5 through 9 of Norman Cook's book *Models of the Atomic Nucleus* [Cook 2010] in an analysis of "Long-Standing Problems," including a chapter on LENR. The impasse goes all the way back to 1933 when the neutron was deemed to be a fundamental particle, just as the proton is fundamental. In SAM, *we consider the neutron to be a proton–electron pair—a combination*, resulting in a nucleus that consists of protons and (inner, or nuclear) electrons. Such a modification of the model for the nucleus represents a major simplification of the physics that has been established around the proton–neutron model. SAM is a new approach to "structure theory" in that it does not force nucleons into a specific lattice. Rather, nature itself has been enlisted to tell SAM which configurations are preferred at the various stages in the growth patterns of the elements. Thus, SAM is an "unforced", natural structural model of the nucleus and, as far as we know, the only such model. This change in the way we view the nucleus is the *first* missing piece of the LENR puzzle.

As discussed in Chapters 11–13 and 15, SAM shows that a nucleus can be more readily altered than commonly assumed. In certain plasma environments, a nucleus can accept protons, electrons, PEPs, and even bigger chunks of nucleons, from elements such as deuterium, helium, carbon, and oxygen nuclei, leading to a nuclear reaction, generally in the form of transmutations, potentially without significant emission of hard radiation.

During our work on SAM, we have identified a heretofore unrecognized source of internal stress (Section 9.5) that is inherent in the structure of heavier nuclei. For certain configurations this stress leads to an unstable geometry that is susceptible to being triggered in a nuclear decay process. We refer to the biggest component of this stress as being caused by "unfulfilled densest packing." Since in most nuclear models the nucleus is a collection of protons and neutrons without organization, questions of structural relations between protons and neutrons do not arise. Hence, this intrinsic stress cannot be found in these models, whereas we are convinced it is a principal factor that enables nuclear fission reactions and therefore some LENR. So, here we have found the *second* missing piece of the LENR puzzle: inherent stress that does not show up in other models of the nucleus.

In Chapter 15 of this book we have shown that *transmutations* can be explained using the insight gained from SAM. Therefore, rather than being a curious by-product of LENR, we view transmutations as being at the core of the LENR phenomenon and mostly responsible for the energy release at nuclear levels. If we are correct then transmutation products should be present in most, if not all, LENR experiments, even if they are not obvious. Nagel and Katinsky [2018] give a broad overview of the state of LENR as of 2018, including results on recent transmutation data. This issue of transmutations of the *cathode* (typically palladium, platinum, or nickel) is not given enough weight, or at least its importance has been downplayed. The appearance of heavier elements such as manganese, cerium, titanium, and ytterbium in relatively simple LENR laboratory experiments is both surprising and problematic because they demand an explanation that is hard to reconcile with current physics models. Since consensus science does not consider transmutations to be possible within a laboratory setting, this conviction is a major obstacle to the validation of LENR. Therefore, the ability to explain transmutations would be the *third* missing piece of the puzzle.

From the beginning, the LENR community has focused on anomalous or excess heat since that is the way LENR manifests primarily. However, occasionally, more violent effects have occurred such as explosions, anomalous heat bursts, vaporization of electrodes, and meltdown of the experiment apparatus (Section 15.6). All of these happened unexpectedly with only modest amounts of input power, typically in the range of milliwatts to watts, sometimes for weeks or months. A detailed account of an "anomalous heat burst" and so-called "heat after death" (HAD) is given in Mizuno's book [Mizuno 1998]. The experiment was running at 24 watts input when the event began to unfold. Mizuno estimated that 114 megajoules of excess heat was produced after the experiment was shut down, but then continued to generate heat for days. These anomalies collectively point to an energy storage mechanism that can apparently remain in effect for a long time.

In this respect, the role of electricity appears to be underappreciated as a causative agent in LENR. Specifically, the concept of breakdown of electrical double-layers capable of storing huge amounts of electrical charge is rarely addressed in the LENR literature.

In our conceptual model for LENR, transmutations take place in electrified environments, through energy release of a double-layer (DL) breakdown, through electric discharges or when plasma transitions into arc mode under the influence of a strong electric field. Plasma breakdown in double-layers has been studied by SAFIRE [Philips et al. 2020, 44].

SAFIRE experienced such an electrical breakdown in 2017 when a tungsten probe for measuring electric fields vaporized in an instant when it got too close to one of the plasma shells (p. 214). The same thing happened in 2019 when the anode melted. Quoting from the same paper:

> *Of particular interest is the formation of visible striations, or 'double layers' (DLs). Space probes have verified the presence of DLs in space, the magnetospheres of Earth, Jupiter, Saturn and in the solar wind. The theory and understanding of DLs, both in our local near-earth environment, as well as in the laboratory, is in its infancy.* [Philips et al. 2020, 44]

On the right of Figure 16.1 are shown several concentric shells of double-layers inside a plasma in the SAFIRE chamber. The images are from the *Safire* website. Extraordinarily strong electric fields were measured before the probe was vaporized. On the left of Figure 16.1 we see anode "tufts" that coalesce in a plasma shell as the current is increased. These double-layers can store an enormous amount of charge/energy. Thus, we identify double-layers as a likely site of this storage mechanism: this is the *fourth* missing piece of the LENR puzzle.

In our model for LENR, we make a connection between the plasma as generated by SAFIRE and the effects taking place in LENR experiments. Seeing that electrical breakdown of double-layers is the trigger initiating a chain reaction of transmutations

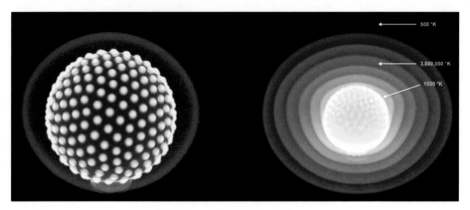

Figure 16.1 SAFIRE double layers.
[Safire Project 2019/science/phase-three.html]

that evaporated the tungsten probe used by SAFIRE in 2017 and caused the anode to melt in 2019, we postulate that the same effect took place on a microscopic scale at the cathode of the Fleischmann and Pons' electrolytic cell. We know that transmutation products have been found in one of the Fleischmann and Pons' cathodes, as described by Biberian. The abstract reads:

> . . . it is shown that silver detected is not due to contamination, but probably to transmutation of palladium by absorption of a deuterium nucleus, because only the ^{107}Ag-isotope is present, and not ^{109}Ag, as in natural silver. Another important result of this study is the determination of the depth at which the reaction occurs: 1.3 microns below the cathode surface. [Biberian 2019]

We take this as confirmation that the original Fleischmann and Pons' cathodes indeed underwent transmutations and that it is reasonable to assume that the Fleischmann and Pons Effect (FPE) is caused by transmutations at the microscopic scale. This phenomenon of electrical breakdown of double-layers would be the *fifth* missing piece of the LENR puzzle.

An example of plausible electrical breakdown on the microscopic scale is shown in Figure 16.2. It is a screenshot of a 1996 infrared video published by P. Mosier-Boss and S. Szpak. The video is discussed by Nagel and Srinivasan who conclude that:

> . . . these temperatures imply energy releases on the scale of nuclear values. The video shows that the small spots turn on and off as time progresses. It is as if there are very local releases of energy at different points of the cathode at different times. [Nagel and Srinivasan 2014]

We hypothesize that these IR "scintillations" on a different type of cathode than that of Fleischman and Pons are microscopic plasmoids being electrically discharged when they encounter a metal conductor. They are the equivalent of the anode tufts from

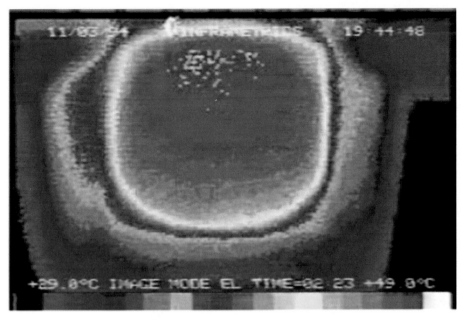

Figure 16.2 Electrical breakdown.
[Szpak et al. 2006, 7]

SAFIRE, most likely being double-layers as well. The spacing of the scintillations is compatible with a micron-size plasmoid:

> *The prospect of understanding the plasma interaction with matter at the submicron to the angstrom level in trapped potential wells in the form of DLs would afford a host of new potential experiments for plasma scientists.* [Philips et al. 2020, 46]

The observations of anomalous heat effect (AHD), "heat after death" (HAD) that characterize the FPE, occur at these scintillations on the cathode surface. In this view, the underlying phenomenon is transmutation of the cathode material in electrical discharges of microscopic plasmoids. In summary, recognition of the equivalence/similarity of SAFIRE results with the microscopic IR scintillations of Mosier-Boss/Szpak is the *sixth* missing piece of the LENR puzzle.

This scenario for LENR is inherently non-linear and unpredictable, characteristic of electrical discharges, and dependent on the local topography of the surface. The discharges modify the surface and change local electric fields through cratering and melting. The assembly of microplasmoids has its own dynamics which is not yet understood or predictable. Energy is stored in the double-layers of plasmoids and can remain there for a long time until something happens that causes the discharge. It is therefore not surprising that the timeframes range from fractions of a second to several days, weeks, or months.

Considering the above effects and phenomena we formulate a phenomenological model for LENR with the following scenario:

a The source of excess energy is provided by interaction of plasma with a conductor/metal in a transmutation chain that is net exothermic. Some reactions may be endothermic using part of the energy released in a previous step or coming from "storage."

b Plasma, consisting of free electrons and hydrogen ions (H+) (or deuterium) can organize itself spontaneously in hollow spherical shells, forming double-layers (DL) of atomic hydrogen and electrons on the scale of microns. These structures can survive for long times, depending on their immediate electrical environment.

c Extremely high electric field values exist close to a DL. Foreign metals introduced into the vicinity of a DL causes electric discharge and breakdown of the DL. This in turn initiates a chain of transmutation events.

d The exact root of transmutations will remain controversial until a true structural model for the nucleus is recognized and used to identify individual reactions. The *Structured Atom Model* is such a model.

Recent developments have allowed the picture of the LENR scenario to be solidified. The research published by the SAFIRE team paints a picture of a surprisingly simple model for the Sun as plasma confined by double-layers. This Sun model has been advocated for over 30 years by the principals of the Electric Universe who maintain that astrophysics, by denying the possibility of electric currents in space, has resulted in a defective theory based on the 80-year-old gravitational model of Eddington.

SAFIRE has made plausible that most of the solar activity takes place in the photosphere, chromosphere, and corona region—not in the solar interior. This finding would have serious consequences for the future of ITER, the International Thermonuclear Experimental Reactor, which is based on the notion that nuclear fusion takes place in the interior of the Sun at a temperature on the order of 15 million Kelvin. Table 16.1 summarizes the foregoing discussion.

16.3 CONCLUSION

In the above discussion on the LENR puzzle and our clues to resolving it, we have implicitly addressed most of the 10 questions put forward by David Nagel [Nagel and Katinsky 2018]. However, what we have described is not a theory, but a heuristic model based on some key observations. Specifically we talked about the notions that a storage mechanism can be present (energy is stored in electrical double layers), that an electrical breakdown occurs, that the breakdown initiates several nuclear processes in an avalanche that ends in the observed transmutations and large amounts of energy are being released through the binding energy difference of the transmutation steps and possibly from the electrical "storage."

Table 16.1 LENR summary discussion.

LENR puzzle pieces	Effect/impact
1. An unforced structural model (SAM) allows nucleons to find "natural" positions (based on proton–electron configurations rather than proton–neutron configurations).	Results in expanded periodic table, duplicating all known elements and isotopes and flagging several "missing" elements.
2. Discovery of internal stress in the nucleus.	Nuclear instability, cause of radioactive decay.
3. Identification of transmutations of electrodes as being common to most LENR experiments.	Plausible explanation of FPE/AHE/HAD (electrodes are the consumable).
4. Identification of plasma double-layers as a likely storage mechanism.	Connection to Ken Shoulders' EVs or EVOs—exotic vacuum objects, also known as condensed (micro-) plasmoids, to ball lightning [Lewis 2009], and to Ohmasa gas (specially treated mixed hydrogen—oxygen gas, containing atomic oxygen, hydrogen, and deuterium).
5. Important role of electrical breakdown of plasma double layers as a trigger to initiate transmutation chain reactions.	Explanation of explosions, heat bursts, meltdowns, runaway heat events.
6. Scalability of plasma from SAFIRE size to microscopic plasmoids.	Recognizing the importance of SAFIRE diagnostics for investigating and understanding LENR.

## CHAPTER 17

The nature of the atom

Now that we have come to the concluding chapter of the book, we can examine the essential differences between the century-old, still popular Bohr model of the atom shown in Figure 1.1 and SAM, as developed in the pages of this book. The figure shows the nucleus as an unstructured collection of positive protons and neutral neutrons, with electrons orbiting at specific distances from the nucleus. The simplistic notion of "orbit" was later changed to the concept of "shells" and "orbitals."

In quantum mechanics an "orbital" is a mathematical function describing the location and wave-like behavior of an outer electron in an atom. Per *Wikipedia*, "this function can be used to calculate the probability of finding any electron in any specific region around the atom's nucleus." [Wikipedia 2021/Atomic_orbital]

The "shell" concept was subsequently applied to the atomic nucleus, under the assumption that nucleons are also organized in shells. Once the mathematics of the wave function was applied to the nucleus, any view of nucleons having fixed locations within the nucleus and thus any notion of nuclear structure, such as in SAM, became impossible to contemplate.

Aside from postulating the existence of structure, the SAM model has a different interpretation of the neutron. We don't recognize the neutron as an independent entity in the nucleus. Instead, we designate the neutron to be a proton that is paired with an electron, or proton–electron pair (PEP), arranged in such a way that electrons act as "glue" between at least two protons. We must recognize that in this concept the so-called "inner electron" or "nuclear electron" is not the same as an outer electron. Nevertheless, we have seen that under certain conditions electrons can move from outer to inner, and vice versa, through β− decay, β+ decay, and electron capture. Another type of electron is the quasi-inner electron, located between endings of branches and the branches themselves (Section 4.4).

This fundamentally changes the picture of the atom: we now have an assembly of positive protons with an electron in between protons, with outer electrons shielding the nucleus and quasi-inner electrons located very close to the proton/electron struc-ture of the nucleus. We initially spoke of the duality between protons and electrons. Without knowing what they are, we can at least deduce how protons and electrons interact. The internal structure of any nucleus is driven by the densest packing princi-ple and, as we have seen, a natural sequence of structures unfolds with each element in the periodic table having its own unique arrangement of protons, inner electrons, quasi-inner electrons, as well as the number and position of the outer electrons. The net positive nucleus has specific points to which outer electrons can form a connection. Each electron is connected to two protons per definition, regardless of the inner, outer, or quasi-inner state.

The outer electrons will shield each atom from each other atom, not dissimilar to the idea of a "force field." In effect, we therefore see a net positive nucleus with a specific structure and a negative sphere around it, leading to a localized "charge separation."

Since the atom is a collection of charges, the idea that we can induce nuclear transmutations through electric forces is the most promising. Also, the idea that this universe is ruled by electric forces that are many magnitudes larger than the gravitational force makes complete sense.

The tendency of the nucleus to grow only at certain spots and in certain directions leads to fractal expansion and branching. As the number of protons increases, branches increasingly start to interfere with each other, building up internal stress. We call this "stress energy," which is stored within the structure. We could say that stress energy in the nucleus is due to unfulfilled densest packing because of its fractal growth pattern. This stored energy ultimately shows up as differences in the SAM binding energy curves compared with the published curves. The structure of the nucleus is in the end also responsible for the observed radioactivity and nuclear decay. What currently is being attributed to the "weak nuclear force" has now at least a partial rational explanation as interaction with the environment, unstable nuclet/ending configurations, and forces applied to endings and branches by quasi-inner electrons.

The confused state of current models of the atom is highlighted in a publication by Nancy L. Bowen [Bowen 2018]. In her blog *New Concepts in Nuclear Physics* she lists the roadblocks to understanding the nuclear force as follows:

- the Schrödinger equation,
- the Copenhagen interpretation of the Heisenberg Uncertainty Principle, and
- spherically shaped nuclides.

Fortunately, with SAM none of these roadblocks even arise as problems because the mere existence of nuclear structure eliminates any quantum mechanics considerations.

This book presents a mostly complete picture of the status to date of the *Structured Atom Model* including the various discoveries we have made so far. We have tried to impart to the reader a sense that it is not necessary to have deep knowledge of textbook nuclear physics to grasp the various properties of the atomic nucleus.

The atom, elements, and the periodic table of elements still have secrets to be uncovered. This book is an attempt to describe the results of a work in progress, based on a fresh new way of looking at the atom with a sharp focus on the nucleus. We have shown connections to known phenomena and observations, some going back centuries as in the case of transmutations and alchemy.

By insisting on maintaining compatibility between SAM and the available experimental data, we think that the model has shown its adaptability and is proving to be a realistic representation of the atom and the nucleus. This iterative process ultimately has resulted in a structure for each element in the periodic table of elements as well as a new PTE layout. That we could do this and at the same time also predict several known but unexplained phenomena cannot be a coincidence.

With this in place, it should be possible to chart all element paths of creation. However that is a task for a later book.

Throughout the book we indicate numerous research topics that need deeper investigation to help further evolve and refine the model, as well as theory and science resulting from it. The model is transparent enough that anyone can pick up on these topics and be the one to make the next discovery. For example, there are many places in the book where we identify missing elements, some unstable and some stable. Those elements should be out there and we just must look for them. Could it be that scientists stopped looking once all the spots in the old PTE were taken? If so, there is still much to discover. And if a few of them are stable (like some of the missing noble gases, maybe with the help of additional PEPs), why did nature skip them? It happens, of course, because nature prefers the building phase over the capping phase. However, it might be possible to create them artificially.

. . . the hunt is on . . .

Appendices

APPENDIX A: THE STANDARD PERIODIC TABLE OF ELEMENTS IN SAM

Appendix A gives a representation of the standard periodic table of elements (PTE) based on SAM. The standard isotope is displayed per element (Standard Model definition) with the primary oxidation states/valence numbers. The lanthanides and actinides are separated out, as is typically done.

The PTE is available on our website https://structuredatom.org/atomizer/pte and allows you to change the cell content. With the available controls you can switch on/off the following properties:

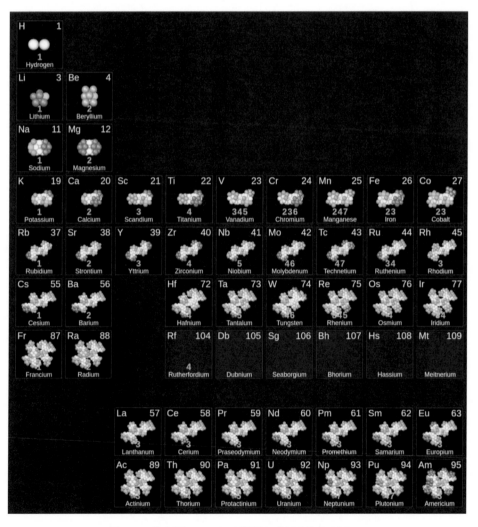

Figure A.1 The standard PTE in SAM (*left side*).

- display large symbol
- display small symbol
- display element name
- display atomic number
- display primary oxidation state large
- display primary oxidation state mall
- display all oxidation states (primary states highlighted)
- display primary image.

A click on one element provides additional information (e.g., isotope abundance percentages).

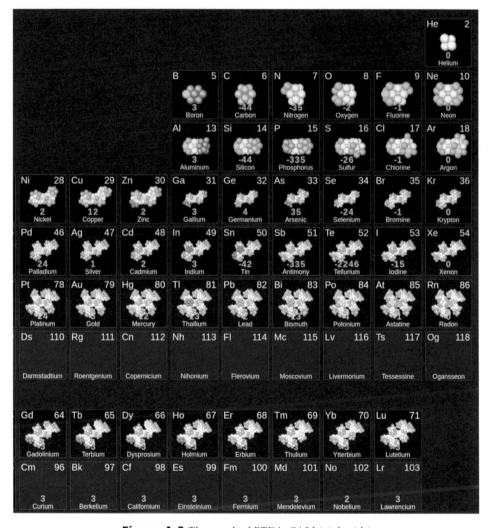

Figure A.2 The standard PTE in SAM (*right side*)

APPENDIX B: THE NEW SAM NUCLEAR PERIODIC TABLE OF ELEMENTS

The new SAM nuclear PTE has 42 columns, by pulling in the lanthanides/actinides into the main sequence and because of new predicted elements. Alkali metals are colored red, alkali-earth metals are colored orange, halogens are colored blue, and

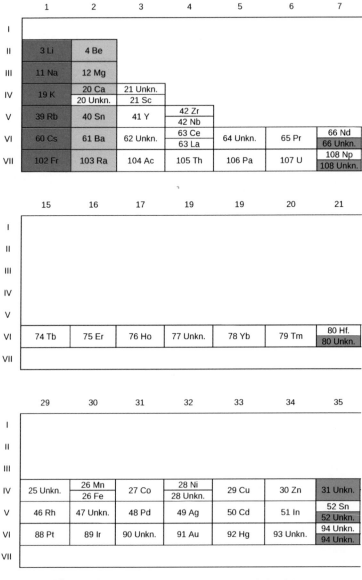

Figure B.1 The new SAM nuclear PTE (*left side*).

noble gases are colored light green. The missing/unknown noble gases are very evenly spaced. Be aware that the number of unknowns is much bigger, here we just covered the empty spots with numbers and what could easily be identified. The new SAM nuclear PTE is by no means complete as for a given element number there could be more structural options and isomeric configurations.

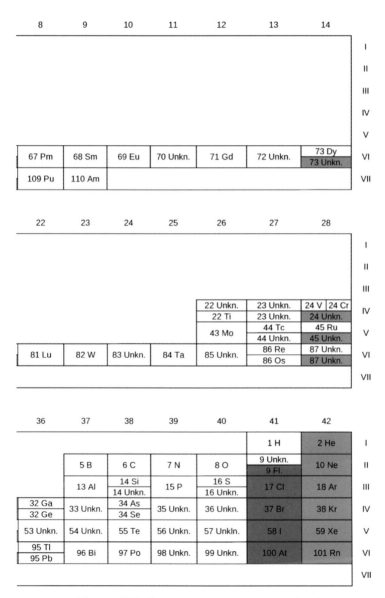

Figure B.2 The new SAM nuclear PTE (*right side*).

APPENDIX C: MAIN POSTULATIONS OF SAM

The setup of the model so far can be summarized as follows:

- Only the electron and proton are fundamental.
- The proton and the electron describe a duality. They are opposites, yet do not annihilate each other.
- Charge-based forces active in the nucleus—acting between protons and electrons and each other are enough to keep the nucleus in a fixed structure.
- The geometry of the nucleus as a result of the forces acting in it follows the "spherical dense packing" principle.
- The atom has a definite organization that is responsible for all the attributes of an element.
- The neutron is not a fundamental particle at the same level as the proton but is redefined as a close connection between a nuclear electron and a proton.
- A stable element has a stable nucleus. There is no movement without cause in the structure of the nucleus. Every proton and every inner electron has a fixed position in the nucleus unless the nucleons become excited.
- The inner structure of the atom (nucleus) determines the outer electron structure (electron shield).

APPENDIX D: MAIN FINDINGS BASED ON SAM

Appendix D provides the main findings in this book based on SAM:

- We identified the basic growth patterns of the nucleus. Growth is a repetition of endings and nuclets creating a carbon nuclet backbone with cappings.
- The nucleus branches and elongates. There is a natural limit to the growth and size of the nucleus.
- There is no island of stability after lead.
- There are three types of electrons: outer, inner, and quasi-inner electrons all of which play different roles.
- Building branches and elongating is a trade-off against the densest packing rules to allow growth. However, this then creates the necessity to store additional energy in the nucleus. This is energy that can be released through fission.
- The structure of the nucleus determines which type of decay can happen.
- The asymmetric breakup of nuclei during fission can be traced to structural properties.
- Proton capture is considered a likely mechanism for inducing nuclear reactions.
- The new numbering system of the PTE and the proposed nuclear structure show that there must be many missing as yet undiscovered elements. Most of them will be unstable, but some are stable.

APPENDIX E: RESEARCH TOPICS

Appendix E presents a list of important research topics:

- What is the relation between inner, quasi-inner, and outer electrons?
- In what way does the inner electron differ from the outer electron?
- In what way does the quasi-inner electron differ from the inner and the outer electron?
- What are the preferred docking points for PEPs? What is the logic behind the order of their positioning?
- Why has nitrogen such a strange electron affinity value?
- How to determine the magnetic moment of a nucleus, especially its direction?
- How to determine the magnetism of an atom from its structure?
- Are the group categorizations of elements of the Standard Model correct, when viewed with the structure of the nucleus in mind?
- How does the number of connections affect the strength of material?
- Develop the Semi-Empirical Binding Energy Formula for SAM.
- Why must the five-ending be completed before further growth is possible?
- More details on why the β decay spectrum is continuous are required.
- Why do we see sometimes 2×0.511 MeV gamma rays at roughly 180° during β+ decay? Is it really the destruction of a deuteron?
- Where does the relation of atomic sizes come from relative to the running cycle-of-eight?
- What is the actual size of the electrons and what does that mean?
- Get more details on how the gamma, x-ray and photon-specific emission lines of decay steps exactly relate to steps of the resettling nucleus.
- How easy is it for a PEP (neutron) connected to a nucleus to connect to a proton and what mechanism is responsible for this?
- Does the direction of impacting particles (PEPs/protons) influence the chance of hitting the right spot?

APPENDIX F: IAEA BINDING ENERGY DATA WITH ADDITIONAL SAM DATA

Table F.1 represents a selection of binding energy data as available on the *IAEA* website. We added the SAM binding energy data for enhancement and comparison.

Table F.1 *SAM line* binding energy versus total binding energy.

Isotope	SAM # lines	SAM line average BE (keV)	Average BE (keV)	SAM line total BE (MeV)	Total BE (MeV)	Stress energy (MeV)
H-2	1	1112.500	1112.283	2.225	2.225	0.000
H-3	4	2966.667	2827.265	8.900	8.482	−0.418
He-3	3	2225.000	2572.680	6.675	7.718	1.043
He-4	12	6675.000	7073.915	26.700	28.296	1.596
Li-6	15	5562.500	5332.331	33.375	31.994	−1.381
Li-7	19	6039.286	5606.439	42.275	39.245	−3.030
Be-9	27	6675.000	6462.668	60.075	58.164	−1.911
B-10	31	6897.500	6475.083	68.975	64.751	−4.224
B-11	35	7079.545	6927.732	75.650	76.205	−0.555
C-12	41	7602.083	7680.144	91.225	92.162	0.937
C-13	44	7530.769	7469.849	97.900	97.108	−0.792
C-14	47	7469.643	7520.319	104.575	105.284	0.709
N-14	49	7787.500	7475.614	109.025	104.659	−4.366
N-15	52	7713.333	7699.460	115.700	115.492	−0.208
O-16	57	7926.563	7976.206	126.825	127.619	0.794
O-17	62	8114.706	7750.728	133.500	131.762	−1.738
O-18	65	8034.722	7767.097	144.625	139.808	−4.817
F-19	70	8197.368	7779.018	155.750	147.801	−7.949
Ne-20	73	8121.250	8032.240	162.425	160.645	−1.780
Ne-21	78	8264.286	7971.713	173.550	167.406	−6.144
Ne-22	83	8394.318	8080.465	184.675	177.770	−6.905
Na-23	86	8319.565	8111.493	191.350	186.564	−4.786
Mg-24	89	8251.042	8260.709	198.025	198.257	0.232
Mg-25	92	8188.000	8223.502	204.700	205.588	0.888
Mg-26	95	8129.808	8333.870	211.375	216.681	5.306
Al-27	102	8405.556	8331.553	226.950	224.952	−1.998
Si-28	107	8502.679	8447.744	238.075	236.537	−1.538
Si-29	110	8439.655	8448.635	244.750	245.010	0.260
Si-30	113	8380.833	8520.654	251.425	255.620	4.195

Isotope	SAM # lines	SAM line average BE (keV)	Average BE (keV)	SAM line total BE (MeV)	Total BE (MeV)	Stress energy (MeV)
P-31	118	8469.355	8481.167	262.550	262.916	0.366
S-32	123	8552.344	8493.129	273.675	271.780	−1.895
S-33	128	8630.303	8497.630	284.800	280.422	−4.378
S-34	131	8572.794	8583.498	291.475	291.839	0.364
Cl-35	136	8645.714	8520.278	302.600	298.210	−4.390
Ar-36	139	8590.972	8519.909	309.275	306.717	−2.558
Cl-37	142	8539.189	8570.281	315.950	317.100	1.150
Ar-38	149	8724.342	8614.280	331.525	327.343	−4.182
K-39	152	8671.795	8557.025	338.200	333.724	−4.476
Ca-40	155	8621.875	8551.303	344.875	342.052	−2.823
K-41	158	8574.390	8576.072	351.550	351.619	0.069
Ca-42	161	8529.167	8616.563	358.225	361.896	3.671
Ca-43	164	8486.047	8600.663	364.900	369.829	4.929
Ca-44	167	8444.886	8658.175	371.575	380.960	9.385
Sc-45	176	8702.222	8618.931	391.600	387.852	−3.748
Ti-46	179	8658.152	8656.451	398.275	398.197	−0.078
Ti-47	182	8615.957	8661.227	404.950	407.078	2.128
Ti-48	185	8575.521	8723.006	411.625	418.704	7.079
Ti-49	188	8536.735	8711.157	418.300	426.847	8.547
Cr-50	195	8677.500	8701.032	433.875	435.052	1.177
V-51	200	8725.490	8742.099	445.000	445.847	0.847
Cr-52	201	8600.481	8775.989	447.225	456.351	9.126
Cr-53	204	8564.151	8760.198	453.900	464.290	10.390
Fe-54	211	8693.981	8736.382	469.475	471.765	2.290
Mn-55	216	8738.182	8765.022	480.600	482.076	1.476
Fe-56	221	8780.804	8790.354	491.725	492.260	0.535
Fe-57	224	8743.860	8770.279	498.400	499.906	1.506
Ni-58	229	8784.914	8732.059	509.525	506.459	−3.066
Co-59	234	8824.576	8768.035	520.650	517.314	−3.336
Ni-60	239	8862.917	8780.774	531.775	526.846	−4.929
Ni-61	242	8827.049	8765.025	538.450	534.667	−3.783
Ni-62	245	8792.339	8794.553	545.125	545.262	0.137
Cu-63	250	8829.365	8752.138	556.250	551.385	−4.865
Zn-64	253	8795.703	8735.905	562.925	559.098	−3.827
Cu-65	260	8900.000	8757.096	578.500	569.211	−9.289
Zn-66	263	8866.288	8759.632	585.175	578.136	−7.039

Isotope	SAM # lines	SAM line average BE (keV)	Average BE (keV)	SAM line total BE (MeV)	Total BE (MeV)	Stress energy (MeV)
Zn-67	266	8833.582	8734.152	591.850	585.188	−6.662
Zn-68	269	8801.838	8755.680	598.525	595.386	−3.139
Ga-69	274	8835.507	8724.579	609.650	601.996	−7.654
Ge-70	279	8868.214	8721.700	620.775	610.519	−10.256
Ga-71	282	8837.324	8717.604	627.450	618.950	−8.500
Ge-72	287	8869.097	8731.745	638.575	628.686	−9.889
Ge-73	290	8839.041	8705.049	645.250	635.469	−9.781
Se-74	295	8869.932	8687.715	656.375	642.891	−13.484
As-75	300	8900.000	8700.874	667.500	652.566	−14.934
Se-76	305	8929.276	8711.477	678.625	662.072	−16.553
Se-77	308	8900.000	8694.690	685.300	669.491	−15.809
Se-78	311	8871.474	8717.806	691.975	679.989	−11.986
Br-79	314	8843.671	8687.594	698.650	686.320	−12.330
Kr-80	317	8816.563	8692.928	705.325	695.434	−9.891
Br-81	324	8900.000	8695.946	720.900	704.372	−16.528
Kr-82	327	8872.866	8710.675	727.575	714.275	−13.300
Kr-83	332	8900.000	8695.729	738.700	721.746	−16.954
Sr-84	333	8820.536	8677.512	740.925	728.911	−12.014
Rb-85	340	8900.000	8697.441	756.500	739.282	−17.218
Sr-86	343	8874.128	8708.456	763.175	748.927	−14.248
Rb-87	350	8951.149	8710.983	778.750	757.856	−20.894
Sr-88	353	8925.284	8732.595	785.425	768.468	−16.957
Y-89	356	8900.000	8713.978	792.100	775.544	−16.556
Zr-90	359	8875.278	8709.969	798.775	783.897	−14.878
Zr-91	362	8851.099	8693.314	805.450	791.092	−14.358
Mo-92	367	8875.815	8657.730	816.575	796.511	−20.064
Nb-93	374	8947.849	8664.184	832.150	805.769	−26.381
Tc-94	375	8876.330	8608.736	834.375	809.221	−25.154
Mo-95	378	8853.158	8648.720	841.050	821.628	−19.422
Ru-96	383	8876.823	8609.412	852.175	826.504	−25.671
Mo-97	384	8808.247	8635.092	854.400	837.604	−16.796
Ru-98	393	8922.704	8620.313	874.425	844.791	−29.634
Ru-99	396	8900.000	8608.712	881.100	852.262	−28.838
Ru-100	401	8922.250	8619.359	892.225	861.936	−30.289
Ru-101	404	8900.000	8601.365	898.900	868.738	−30.162
Ru-102	407	8878.186	8607.427	905.575	877.958	−27.617

Isotope	SAM # lines	SAM line average BE (keV)	Average BE (keV)	SAM line total BE (MeV)	Total BE (MeV)	Stress energy (MeV)
Rh-103	416	8986.408	8584.192	925.600	884.172	−41.428
Pd-104	417	8921.394	8584.848	927.825	892.824	−35.001
Pd-105	422	8942.381	8570.650	938.950	899.918	−39.032
Pd-106	427	8962.972	8579.992	950.075	909.479	−40.596
Ag-107	430	8941.589	8553.900	956.750	915.267	−41.483
Cd-108	433	8920.602	8550.019	963.425	923.402	−40.023
Ag-109	440	8981.651	8547.915	979.000	931.723	−47.277
Pd-110	443	8960.682	8547.162	985.675	940.188	−45.487
Cd-111	448	8980.180	8537.079	996.800	947.616	−49.184
Cd-112	453	8999.330	8544.730	1007.925	957.010	−50.915
In-113	456	8978.761	8522.929	1014.600	963.091	−51.509
Sn-114	459	8958.553	8522.566	1021.275	971.573	−49.702
Sn-115	464	8977.391	8514.069	1032.400	979.118	−53.282
Sn-116	467	8957.543	8523.116	1039.075	988.681	−50.394
Sn-117	470	8938.034	8509.611	1045.750	995.624	−50.126
Sn-118	473	8918.856	8516.533	1052.425	1004.951	−47.474
Sn-119	476	8900.000	8499.449	1059.100	1011.434	−47.666
Sn-120	479	8881.458	8504.492	1065.775	1020.539	−45.236
Sb-121	486	8936.777	8482.066	1081.350	1026.330	−55.020
Te-122	491	8954.713	8478.140	1092.475	1034.333	−58.142
Te-123	494	8936.179	8465.546	1099.150	1041.262	−57.888
Te-124	497	8917.944	8473.279	1105.825	1050.687	−55.138
Te-125	500	8900.000	8458.045	1112.500	1057.256	−55.244
Te-126	503	8882.341	8463.248	1119.175	1066.369	−52.806
I-127	512	8970.079	8445.487	1139.200	1072.577	−66.623
Xe-128	515	8952.148	8443.298	1145.875	1080.742	−65.133
Xe-129	520	8968.992	8431.390	1157.000	1087.649	−69.351
Xe-130	525	8985.577	8437.731	1168.125	1096.905	−71.220
Xe-131	530	9001.908	8423.736	1179.250	1103.509	−75.741
Ba-132	531	8950.568	8409.375	1181.475	1110.038	−71.438
Cs-133	538	9000.376	8409.978	1197.050	1118.527	−78.523
Ba-134	541	8983.022	8408.171	1203.725	1126.695	−77.030
Ba-135	546	8998.889	8397.533	1214.850	1133.667	−81.183
Ce-136	547	8949.081	8373.760	1217.075	1138.831	−78.244
Ba-137	554	8997.445	8391.827	1232.650	1149.680	−82.970
Ba-138	557	8980.616	8393.420	1239.325	1158.292	−81.033

Isotope	SAM # lines	SAM line average BE (keV)	Average BE (keV)	SAM line total BE (MeV)	Total BE (MeV)	Stress energy (MeV)
La-139	562	8996.043	8378.025	1250.450	1164.545	−85.905
Ce-140	567	9011.250	8376.317	1261.575	1172.684	−88.891
Pr-141	568	8963.121	8353.992	1263.800	1177.913	−85.887
Nd-142	571	8947.007	8346.030	1270.475	1185.136	−85.339
Nd-143	576	8962.238	8330.488	1281.600	1191.260	−90.340
Sm-144	577	8915.451	8303.679	1283.825	1195.730	−88.095
Nd-145	586	8992.069	8309.187	1303.850	1204.832	−99.018
Nd-146	591	9006.678	8304.092	1314.975	1212.397	−102.578
Pm-147	594	8990.816	8284.372	1321.650	1217.803	−103.847
Sm-148	597	8975.169	8279.633	1328.325	1225.386	−102.939
Sm-149	602	8989.597	8263.466	1339.450	1231.256	−108.194
Sm-150	607	9003.833	8261.621	1350.575	1239.243	−111.332
Eu-151	610	8988.411	8239.297	1357.250	1244.134	−113.116
Gd-152	611	8943.914	8233.401	1359.475	1251.477	−107.998
Eu-153	620	9016.340	8228.699	1379.500	1258.991	−120.509
Gd-154	621	8972.240	8224.796	1381.725	1266.619	−115.106
Gd-155	624	8957.419	8213.251	1388.400	1273.054	−115.346
Dy-156	627	8942.788	8192.433	1395.075	1278.020	−117.055
Gd-157	634	8985.032	8203.504	1410.650	1287.950	−122.700
Gd-158	641	9026.741	8201.819	1426.225	1295.887	−130.338
Tb-159	640	8955.975	8188.800	1424.000	1302.019	−121.981
Gd-160	647	8997.344	8183.014	1439.575	1309.282	−130.293
Dy-161	652	9010.559	8173.310	1450.700	1315.903	−134.797
Er-162	653	8968.673	8152.397	1452.925	1320.688	−132.237
Dy-163	662	9036.503	8161.785	1472.950	1330.371	−142.579
Dy-164	665	9022.104	8158.714	1479.625	1338.029	−141.596
Ho-165	666	8980.909	8146.964	1481.850	1344.249	−137.601
Er-166	673	9020.633	8141.959	1497.425	1351.565	−145.860
Er-167	678	9033.234	8131.746	1508.550	1358.002	−150.548
Yb-168	677	8966.220	8111.898	1506.325	1362.799	−143.526
Tm-169	680	8952.663	8114.473	1513.000	1371.346	−141.654
Er-170	689	9017.794	8111.959	1533.025	1379.033	−153.992
Yb-171	690	8978.070	8097.882	1535.250	1384.738	−150.512
Yb-172	697	9016.424	8097.429	1550.825	1392.758	−158.067
Yb-173	702	9028.613	8087.427	1561.950	1399.125	−162.825
Hf-174	703	8989.511	8068.533	1564.175	1403.925	−160.250

Isotope	SAM # lines	SAM line average BE (keV)	Average BE (keV)	SAM line total BE (MeV)	Total BE (MeV)	Stress energy (MeV)
Lu-175	706	8976.286	8069.140	1570.850	1412.100	−158.751
Hf-176	713	9013.778	8061.359	1586.425	1418.799	−167.626
Hf-177	718	9025.706	8051.835	1597.550	1425.175	−172.375
Hf-178	723	9037.500	8049.442	1608.675	1432.801	−175.874
Hf-179	728	9049.162	8038.546	1619.800	1438.900	−180.900
Hf-180	731	9035.972	8034.930	1626.475	1446.287	−180.188
Ta-181	730	8973.757	8023.400	1624.250	1452.235	−172.015
W-182	737	9010.027	8018.308	1639.825	1459.332	−180.493
W-183	740	8997.268	8008.322	1646.500	1465.523	−180.977
W-184	743	8984.647	8005.077	1653.175	1472.934	−180.241
Re-185	746	8972.162	7991.009	1659.850	1478.337	−181.513
Os-186	751	8983.737	7982.831	1670.975	1484.807	−186.168
Os-187	756	8995.187	7973.780	1682.100	1491.097	−191.003
Os-188	761	9006.516	7973.864	1693.225	1499.086	−194.139
Os-189	766	9017.725	7963.002	1704.350	1505.007	−199.343
Os-190	771	9028.816	7962.104	1715.475	1512.800	−202.675
Ir-191	770	8969.895	7948.113	1713.250	1518.090	−195.160
Pt-192	777	9004.297	7942.491	1728.825	1524.958	−203.867
Ir-193	780	8992.228	7938.133	1735.500	1532.060	−203.440
Pt-194	787	9026.160	7935.941	1751.075	1539.573	−211.502
Pt-195	792	9036.923	7926.552	1762.200	1545.678	−216.522
Pt-196	797	9047.577	7926.529	1773.325	1553.600	−219.725
Au-197	796	8990.355	7915.654	1771.100	1559.384	−211.716
Hg-198	799	8978.662	7911.552	1777.775	1566.487	−211.288
Hg-199	804	8989.447	7905.279	1788.900	1573.151	−215.749
Hg-200	809	9000.125	7905.895	1800.025	1581.179	−218.846
Hg-201	814	9010.697	7897.560	1811.150	1587.410	−223.740
Hg-202	817	8999.134	7896.850	1817.825	1595.164	−222.661
Tl-203	818	8965.764	7886.053	1820.050	1600.869	−219.181
Pb-204	823	8976.348	7879.932	1831.175	1607.506	−223.669
Tl-205	826	8965.122	7878.394	1837.850	1615.071	−222.779
Pb-206	833	8997.209	7875.362	1853.425	1622.325	−231.100
Pb-207	838	9007.488	7869.866	1864.550	1629.062	−235.488
Pb-208	843	9017.668	7867.453	1875.675	1636.430	−239.245
Bi-209	846	9006.459	7847.987	1882.350	1640.229	−242.121
Po-210	849	8995.357	7834.346	1889.025	1645.213	−243.812

Isotope	SAM # lines	SAM line average BE (keV)	Average BE (keV)	SAM line total BE (MeV)	Total BE (MeV)	Stress energy (MeV)
Po-211	854	9005.450	7818.784	1900.150	1649.763	−250.387
Po-212	859	9015.448	7810.243	1911.275	1655.772	−255.503
Bi-213	866	9046.244	7791.021	1926.850	1659.487	−267.363
At-214	863	8972.780	7776.366	1920.175	1664.142	−256.033
Rn-215	866	8962.093	7763.814	1926.850	1669.220	−257.630
Po-216	879	9054.514	7758.819	1955.775	1675.905	−279.870
Po-217	884	9064.055	7741.360	1966.900	1679.875	−287.025
Rn-218	881	8991.858	7738.752	1960.225	1687.048	−273.177
Rn-219	886	9001.598	7723.777	1971.350	1691.507	−279.843
Rn-220	891	9011.250	7717.254	1982.475	1697.796	−284.679
Rn-221	896	9020.814	7701.393	1993.600	1702.008	−291.592
Rn-222	901	9030.293	7694.497	2004.725	1708.178	−296.547
Fr-223	904	9019.731	7683.664	2011.400	1713.457	−297.943
Ra-224	907	9009.263	7679.922	2018.075	1720.303	−297.772
Fr-225	914	9038.444	7662.940	2033.650	1724.162	−309.489
Ra-226	917	9027.987	7661.962	2040.325	1731.603	−308.722
Ac-227	920	9017.621	7650.707	2047.000	1736.710	−310.290
Ra-228	927	9046.382	7642.428	2062.575	1742.474	−320.101
Ra-229	932	9055.459	7628.485	2073.700	1746.923	−326.777
Th-230	933	9025.761	7630.996	2075.925	1755.129	−320.796
Pa-231	936	9015.584	7618.426	2082.600	1759.856	−322.744
Th-232	943	9043.858	7615.033	2098.175	1766.688	−331.487
U-233	944	9014.592	7603.956	2100.400	1771.722	−328.678
U-234	949	9023.611	7600.715	2111.525	1778.567	−332.958
U-235	954	9032.553	7590.914	2122.650	1783.865	−338.785
Np-236	957	9022.564	7579.214	2129.325	1788.695	−340.630
Np-237	962	9031.435	7574.989	2140.450	1795.272	−345.178
Pu-238	965	9021.534	7568.360	2147.125	1801.270	−345.855
Pu-239	970	9030.335	7560.318	2158.250	1806.916	−351.334
Pu-240	973	9020.521	7556.042	2164.925	1813.450	−351.475
Am-241	978	9029.253	7543.278	2176.050	1817.930	−358.120

The stress energy column clearly shows the divergence between SAM calculations and observed values. This divergence is explained through the application of only 1st-order and 2nd-order organizational patterns in the SAM numbers.

APPENDIX G: TOOLS

The development of SAM would not have been possible without some tools.

G.1 MARBLES

To figure out spherical dense packing you can glue marbles together.

G.2 MAGNETS

Spherical magnets also work up to a degree for spherical dense packing, but they are dipoles, so there are deficiencies in the buildup. However, big structures can be created with magnets as Figure G.1 proves—an early attempt at the backbone structure.

G.3 ATOM-VIEWER

Atom-Viewer is a tool written in Javascript to view the nucleus of any atom according to SAM. You can find the Atom-Viewer on the *Structured-Atom* website under https://structuredatom.org/atomizer/atom-viewer.

Figure G.1 Early backbone test with magnets.

G.4 THE CHEMICAL PERIODIC TABLE OF ELEMENTS

The *Structured-Atom* website provides a view of the original periodic table of elements in SAM under https://structuredatom.org/atomizer/pte. It is also printed in this book in Appendix A. The new/missing/unknown elements are not shown there.

G.5 ATOM-BUILDER

Atom-Builder is a tool not available to the public. It allows creating and changing the elements and isotopes you see in Atom-Viewer.

G.6 MOLECULE-BUILDER

Molecule-Builder will be the logical extension to the Atom-Builder at the molecule level. It does not as yet exist.

APPENDIX H: THE ELEMENTS AND THEIR ISOTOPES

Table H.1 lists the explanation and calculation rules for some of the fields in the reference tables per isotope that were moved to the *Structured Atom* website. We cover in this reference common isotopes and those mentioned in the book.

Table H.1 Reference explanation/calculation.

Field	Explanation/Calculation
Number of deuterons	Determined by the structure (nuclets, endings)
Number of single carbon nuclet protons	Determined by the super-structure: one for each carbon nuclet with 11 protons (= all except the first one, which has 12 protons)
Number of quasi-inner electrons	Determined by the structure: space between nuclets and between branches provides possible spots
Number of additional required proton–electron pairs	Determined by the structure: PEPs required for stability
Number of additional gap proton–electron pairs	Determined by the structure: PEPs not required, but filling a gap ("neutron" gap)
Number of additional other proton–electron pairs	Determined by the structure: all other PEPs not required and not filling a gap
Total number of protons in the nucleus	Twice the number of deuterons + number of single carbon nuclet protons + number of additional required proton–electron pairs + number of additional gap proton–electron pairs + number of additional other proton–electron pairs
Number of outer electrons	Number of deuterons + number of single carbon nuclet protons - number of quasi-inner electrons
Number of inner electrons	Number of deuterons + number of additional required proton–electron pairs + number of additional gap proton–electron pairs + number of additional other proton–electron pairs
Element-/Atomic number	Number of deuterons + number of single carbon nuclet protons unable to pull in a quasi-inner electron

Table H.2 lists the deuteron count as well as the single proton count of a nuclet/ending:

Table H.2 Deuteron, single proton, and PEP count of nuclets/endings.

Nuclet/Ending	Deuteron count	Single proton count	PEP count
First carbon nuclet	6	0	0
Other carbon nuclet	5	1	0
Boron ending	5	0	0
Beryllium ending	4	0	0
Lithium nuclet	3	0	0
Five-ending	2	0	1
Four-ending	2	0	0
Two-ending	1	0	0

The fields

- Group
- Magnetic dipole moment
- Spin

listed in the reference (which was moved to the website) we consider to be legacy information, that needs to be redone in the future.

Information specific to SAM consists of

- SAM lines
- SAM line nucleus BE
- MBS radius (based on proton radius 1)
- MBS Vol./#p.

To reduce the extent of the book we moved the element/isotope reference to the *Structured Atom* website. You can find the reference at https://structuredatom.org/book/appendix_h.pdf.

SCAN ME

APPENDIX I: OUTER ELECTRON VIEW OF THE ELEMENTS

After having developed the new SAM nuclear PTE we can look back at the outer electron count and make a comparison with the new element count.

Table I.1 Categorizing elements by number of outer electrons.

# Outer electrons	Elements
1	001 Hydrogen
2	002 Helium
3	003 Lithium
4	004 Beryllium
5	005 Boron
6	006 Carbon
7	007 Nitrogen
8	008 Oxygen
9	009 Fluorine
10	010 Neon
11	011 Sodium
12	012 Magnesium
13	013 Aluminum
14	014 Silicon, 014 Missing element 29
15	015 Phosphorus
16	016 Sulfur, 016 Missing element 32
17	017 Chlorine
18	018 Argon
19	019 Potassium
20	020 Calcium, 020 Missing element 41
21	021 Missing element 43, 021 Scandium
22	022 Missing element 45, 022 Titanium
23	023 Missing element 49, 023 Missing element 50, 024 Vanadium
24	024 Chromium, 024 Missing element 50
25	025 Missing element 53, 025 Manganese
26	026 Iron
27	027 Cobalt
28	028 Nickel, 028 Missing element 62
29	029 Copper
30	030 Zinc
31	031 Missing element 66, 032 Gallium

# Outer electrons	Elements
32	032 Germanium, 033 Missing element 70
33	034 Arsenic
34	034 Selenium
35	035 Missing element 75, 037 Bromine
36	036 Missing element 78, 038 Krypton
37	039 Rubidium
38	040 Strontium
39	041 Yttrium
40	042 Zirconium
41	042 Niobium
42	043 Molybdenum
43	044 Technetium, 044 Missing element 95
44	045 Ruthenium, 045 Missing element 98
45	046 Rhodium, 047 Missing element 108
46	048 Palladium
47	049 Silver
48	050 Cadmium
49	051 Indium
50	052 Tin, 052 Missing element 52
51	053 Missing element 116, 054 Missing element 119, 056 Antimony
52	054 Missing element 118, 055 Tellurium, 057 Missing element 125
53	058 Iodine
54	059 Xenon
55	060 Cesium
56	061 Barium
57	062 Missing element 133, 063 Lanthanum
58	063 Cerium, 064 Missing element 139
59	065 Praseodymium
60	066 Neodymium, 066 Missing element 142
61	067 Promethium
62	068 Samarium
63	069 Europium
64	070 Missing element 151, 071 Gadolinium
65	072 Missing element 154, 074 Terbium
66	073 Dysprosium, 073 Missing element 156
67	076 Holmium
68	075 Erbium, 077 Missing element 165

# Outer electrons	Elements
69	079 Thulium
70	078 Ytterbium, 080 Missing element 170
71	081 Lutetium
72	080 Hafnium
73	084 Tantalum
74	082 Tungsten, 085 Missing element 181
75	083 Missing element 181, 086 Rhenium
76	086 Osmium, 087 Missing element 186
77	087 Missing element 192, 089 Iridium
78	088 Platinum, 090 Missing element 195
79	091 Gold
80	092 Mercury
81	093 Missing element 199, 095 Thallium
82	094 Missing element 201, 094 Missing element 202, 095 Lead
83	096 Bismuth
84	097 Polonium
85	098 Missing element 210, 100 Astatine
86	099 Missing element 212, 101 Radon
87	102 Francium
88	103 Radium
89	104 Actinium
90	105 Thorium
91	106 Protactinium
92	107 Uranium
93	108 Neptunium, 108 Missing element 232
94	109 Plutonium
95	110 Americium

APPENDIX J: BIBLIOGRAPHY

Aureon Energy (2021), *The SAFIRE Project. Transformative Technology*, https://aureon.ca

Beaudette, C (2000), *Excess Heat*, Oak Grove Press, South Bristol, Maine.

Biberian, J. P. (2019), Anomalous Isotopic Distribution of Silver in a Palladium Cathode, *Journal of Condensed Matter Nuclear Science* 29, 211– 218.

Bowen, N. L. (2018), *The Electromagnetic Considerations of the Nuclear Force*, https://newconceptsinnuclearphysics.com/wp-content/uploads/2018/08/The-Electromagnetic-Considerations-of-the-Nuclear-Force-Current.pdf

Encyclopedia Britannica (2021), various articles, https://www.britannica.com

Cook, N. D. (2010), *Models of the Atomic Nucleus*, 2nd edition, Springer Verlag, Heidelberg.

Cook, N. D. & Dallacasa, V. (2014), LENR and Nuclear Structure Theory, *Journal of Condensed Matter Nuclear Sciences* 13, 68–79.

Cook, N. D. & Di Sia, P. (2018), *The "Renaissance," Nuclear Theory*, presented at *ICCF-21*, June, https://www.youtube.com/watch?v=yYzrBtU_Bis

Dash, J. (2003), *Processing Radioactive Materials with Hydrogen Isotope Nuclei*, https://patents.google.com/patent/WO2003098640A2

Filippov, D. V., Rukhadze, A. A., & Urutskoev, L. I. (2006), *Effects of Atomic Electrons on Nuclear Stability and Radioactive Decay*, https://www.lenr-canr.org/acrobat/FilippovDeffectsofa.pdf

Fischer, K., Gärtner, B., Herrmann, T., Hoffmann, M., & Schönherr, S. (2020), Bounding Volumes, *CGAL User and Reference Manual*. CGAL Editorial Board, 5.0.2 edition.

Gurevich, A. V., Antonova, V. P., Chubenko, A. P., Karashtin, A. N., & Mitko, G. G. (2012), Strong Flux of Low-Energy Neutrons Produced by Thunderstorms, *Physical Review Letters* 108, 125001.

Hagelstein, P. (1992), *Summary of The Third Annual Conference on Cold Fusion* 27, http://www.newenergytimes.com/v2/conferences/1992/ICCF3/1992-ICCF3-HagelsteinSummary.pdf

Hodgson, P. E., Gadioli, E., & Gadioli-Erba, E. (1997), *Introductory Nuclear Physics*, Oxford Science Publications, Oxford.

IAEA (2021), *Live Chart of Nuclides*, https://www-nds.iaea.org/relnsd/vcharthtml/VChartHTML.html

Johnson, B. & Johnson, J. (2011), *The Essential Guide to the Electric Universe*, https://www.thunderbolts.info/wp/eu-guides/eg-contents/

Karjono, A. (2012), Own work based on: *Electron Affinities of the Elements 2.png by Sandbh.*, CC-BY-SA 3.0, https://commons.wikimedia.org/w/index.php?curid=22542942

Kervran, C. L. (1980), *Biological Transmutations*, Beekman Publishers Inc., Woodstock.

Krivit, S. B. (2012), Special Report: Cold Fusion is Neither, *New Energy Times Magazine*, July 30, http://www.newenergytimes.com/v2/news/2010/35/ColdFusionisNeither.pdf

Lerner, E. J. (1991), *The Big Bang Never Happened*. Times Books, New York.

Lewis, E. H. (2009), Tracks of Ball Lightning in Apparatus?, *Journal of Condensed Matter Nuclear Sciences* 2, 13-32.

Leybourne, B. (2017), Hurricane Irma 2017: relationships with lightning, gravity, and earthquakes, *Systemics, Cybernetics and Informatics* 5, Number 3, 7–13.

Mizuno, Tadahiko (1998), *Nuclear Transmutation: The Reality of Cold Fusion*, Infinite Energy Press, Concord, New Hampshire.

Nagel, D. J. (2018), Expectations of LENR Theories, *Journal of Condensed Matter Nuclear Science*, 26, 15–31

Nagel, D., J. & Katinsky, S. B. (2018), Overview of ICCF-21 2018, *Infinite Energy*, Issue 141, 11-40, http://www.infinite-energy.com/iemagazine/issue141/ICCF21.pdf

Nagel, D. J. & Srinivasan, M. (2014), Evidence from LENR Experiments for Bursts of Heat, Sound, EM-Radiation and Particles and for Microexplosions, *Journal of Condensed Matter Nuclear Science* 13, 443-454

NIST (2018), *Codata*, https://physics.nist.gov/cgi-bin/cuu/Value?rp|search_for=

Otte, A. (2019), *Electric Universe UK 2019 – Dynamic Earth*, http://www.xn-zeiten-sprnge-llb.de/?p=506; dated 07/17/2019

Philips, C., Childs, M., Clarage, M., & Anderson, P. E. (2020), Review of the SAFIRE Project, *Chronology & Catastrophism Review* 1, 43–47.

Plasma Universe (2021), *Plasma Universe Resources*, https://www.plasma-universe.com/plasma-universe-resources/

PTE (2021), *The Photographic Periodic Table*, https://periodictable.com

Robitaille, P. M. (2011), Liquid metallic hydrogen: a building block for the liquid Sun, *Progress in Physics* 3, 60–74.

SAFIRE Project (2019), *SAFIRE*, https://safireproject.com

Swartz, K. N. (2017), *Metal–Oxygen Fusion Reactor*, https://www.freepatentsonline.com/20170117066.pdf

Szpak, S. Mosier-Boss, P. A., & Gordon, F. E. (2006), *Experimental Evidence for LENR in a Polarized Pd/D Lattice*, presented at *NDIA 2006 Naval S&T Partnership Conference Washington, DC*, https://www.lenr-canr.org/acrobat/SzpakSexperiment.pdf

Thornhill, W. (2010), *Our Misunderstood Sun*, https://www.holoscience.com/wp/our-misunderstood-sun/

Thornhill, W. (2021), *Stars Aren't What You think. The Universe is Electric*, https://www.youtube.com/watch?v=8mdiHUQRKV4&feature=youtu.be

Thornhill, W. & Talbott, D. (2007), *The Electric Universe*, Mikamar Publishing, Portland

ThoughtCo (2021), element composition of the Sun, https://www.thoughtco.com

Urutskoev, L. I., Lksonov, V. I., & Tsinoev, V. G. (2002), Observation of transformation of chemical elements during electric discharge, *Annales Fondation Louis de Broglie* 27, no. 4, 701–726, https://aflb.minesparis.psl.eu/AFLB-274/aflb274p701.pdf

Wikipedia (2021), various articles, https://en.wikipedia.org/wiki

APPENDIX K: GLOSSARY OF TERMS

Anode   The positive side of two electrodes.

Atom   A nucleus and its outer electrons.

Backbone nuclet   A carbon nuclet with complete carbon nuclets on both growth points.

Base structure isomer   Same number of PEPs, same number of single protons plus deuterons, same overall ending types and ending positions, but different PEP positions and energy levels.

BE   *See* Binding energy.

Beryllium ending   A stable, geometrically arranged cluster of 8 protons and 4 inner electrons (4 deuterons) occupying a growth point on a carbon nuclet.

Beta (β) decay   The natural decay of unstable isotopes into isotopes of another element by changing an electron from the inner state to the outer state ($\beta-$) or vice versa ($\beta+$), thereby changing a PEP into a proton or vice versa.

Binding energy   The nuclear binding energy is the minimum energy required to break up a nucleus into its fundamental components, thereby losing structure and deuterons.

Boron ending   A stable, geometrically arranged cluster of 10 protons and 5 inner electrons (5 deuterons) occupying a growth point on a carbon nuclet.

Branch   An outcrop from the backbone nuclets.

Building block   A building block is a two-, four-, five-, lithium, beryllium, boron, or carbon ending.

Building phase   A phase in the growth pattern of the growth points of a carbon nuclet where at least one growth point moved beyond the capping phase.

Capping   A capping is a geometrically arranged cluster of protons and inner electrons that belong to the capping phase. Each capping is also an ending.

Capping phase   A phase in the growth pattern of the growth points of a carbon nuclet where only two- and four-endings or five-endings are added. The capping phase ends when all growth points are covered by four- or five-endings.

Carbon ending   *See* Carbon nuclet.

Carbon nuclet   A slightly distorted icosahedron with two growth points, usually connected to another carbon nuclet through a shared proton.

Cathode   The negative side of two electrodes.

CMNS   *See* Condensed Matter Nuclear Science.

Cold fusion   The concept that fusion can take place at relative benign temperature levels instead of millions of Kelvin.

Component isomer   Same number of PEPs, dame number of single protons plus deuterons, but different ending types and ending positions.

Condensed Matter Nuclear Science   *See* LENR.

Deuterium   The isotope hydrogen-2. It contains the deuteron as nucleus and has one outer electron.

Deuteron   The nucleus of deuterium (hydrogen-2) consists of two protons and one inner electron between those two protons.

Double-layer   A structure in a plasma consisting of two parallel layers of opposite electrical charge. The sheets of charge produce localized excursions of electric potential, resulting in a relatively strong electric field.

Electrolysis   The use of two electrodes (metals) and an electrolyte (water solution) in an electric system.

Electron   A negatively charged "something" forming a duality with the proton.

Element   An element is characterized by the shape of its nucleus for a given number of the deuterons and single protons (protons not bound as part of a deuteron) it contains.

Element number   The element number of an element is the number of deuterons plus the number of single protons unable to pull in outer electrons as quasi-inner electrons into its nucleus. The element number does not hold any information about the shape of the nucleus nor does it uniquely define an element.

Ending   An ending is a stable, geometrically arranged cluster of protons and inner electrons, also called a building block. *See also* nuclet and capping.

Ending type isomer   Same number of PEPs, same number of single protons plus deuterons, same ending types, but different ending positions.

Final ending   *See* Backbone nuclet.

Fission   The splitting of the nucleus into multiple smaller nuclei, thereby potentially releasing stored stress energy.

Five-ending   A four-ending with an additional PEP filling the PEP gap.

Four-ending   Two two-endings arranged in a V-shaped structure on one growth point of a carbon nuclet. Think of it as a squashed tetrahedron structure.

Fusion   The merging of two nuclei creating a heavier element.

Growth point   Each carbon nuclet provides two possible growth points where the nucleus can grow new endings.

Halo-neutron   *See* Halo PEP.

Halo PEP   A PEP not really attached to the proton structure of the nucleus, but floating close to it.

Icosahedron   One of the platonic solids with 12 vertices and 20 faces.

Initial ending   Both growth points of a carbon nuclet being capped with at least a four-ending.

Inner electron   An electron that is located between the protons of the nucleus, acting like "glue".

Isomer   There are three types of isomers: *component, ending type,* and *base structure.*

Isotope   Structure of the same element base, but with a different number of PEPs.

LENR   Low Energy (induced) Nuclear Reactions, sometimes referred to as "Cold Fusion," although that is a more specific term.

Lithium ending   *See* Lithium nuclet.

Lithium nuclet   A stable, geometrically arranged cluster of 6 protons and 3 inner electrons (3 deuterons) occupying a growth point on a carbon nuclet.

Mass   Mass is a quantitative measure of inertia, a fundamental property of all matter. It is, in effect, the resistance that a body of matter offers to a change in its speed or position upon the application of a force.

Mass defect   The difference between the theoretical mass and the actual, measured mass.

MBS *See* Minimal bounding sphere.

MeV Megaelectron volt. The "standard" notation for energy values when dealing with nuclear physics.

Minimal bounding sphere The smallest sphere surrounding a set of given geometrically arranged spheres.

Neutral ending One growth point of a carbon nuclet being capped with at least a four-ending.

Neutron *See* Proton–electron pair.

Neutron gap *See* PEP gap.

Nuclear electron *See* inner electron.

Nucleon In SAM synonymous with the proton, as there are no neutrons as fundamental particles. An argument could be made to include the inner electrons in the nucleon definition, but since we do not know enough yet we decided for now against it.

Nuclet A nuclet is a densely packed, geometrically arranged cluster of protons and inner electrons that belongs to the building phase. Each nuclet is also an ending.

Nucleus The nucleus of an atom is comprised of protons and inner electrons. The quasi-inner electrons must be counted as part of the nucleus too, even if they are not between the protons of the proton structure.

Outer electron An electron positioned far outside the nucleus.

Pentagonal bi-pyramid A geometric structure with 10 triangular faces and many of the attributes of a platonic solid.

PEP *See* Proton–electron pair.

PEP gap A gap in the proton/inner-electron structure of the nucleus, ready to integrate a PEP.

Plasma A collection of atoms in an environment whereby the atoms are in an excited individual (not bound) state and one that is highly conductive in nature.

Platonic solid A platonic solid is constructed from polygonal faces, identical in shape and size as well as having equal sides and angles and with the same number of faces at each vertex.

Proton A positively charged particle, most likely in spherical form. All protons have the same properties and cannot occupy the same space.

Proton capture The concept that the nucleus is sometimes readily able to capture a proton in its nucleus due to active electric forces and/or ionization leading to transmutations (nuclear reactions). From the perspective of SAM this is the primary suspected mechanism triggering LENR.

Proton–electron pair The proton–electron pair replaces the "neutron." It is not an fundamental particle at the same level as the proton.

Quasi-inner electron An outer electron that has been pulled between branches or between sufficiently developed endings on a carbon nuclet.

Radiation Collective term typically used for three different types of energetic output: *alpha*—a helium-4 nucleus (matter); *beta*—electrons (high speed); *gamma*—electromagnetic wave with several keV up to around 8 MeV.

SAM *See* Structured Atom Model.

SAM lines Each touching point or connection between two protons in the nucleus is considered a "line" in SAM. The total number of lines in a nucleus is determined by adding the number of lines corresponding to the individual endings and base carbons.

SAM line binding energy   The binding energy calculated based on a subset of the organizational patterns of SAM. Each SAM line has the assigned value of 2.225 MeV binding energy.

Single proton   A proton that not is part of a PEP and not bound in a deuteron.

Spherical dense packing   The main principle defining how protons cluster when the inner electrons pull them.

Stress energy   The difference between the actual measured/calculated binding energy and the SAM line binding energy. This is energy stored in the nucleus, that can at least partially be released through nuclear reactions.

Structured Atom Model   The new model for the atom (the subject of the whole book).

Tetrahedron   One of the platonic solids; it has 4 vertices and 4 faces.

Tritium   An isotope of hydrogen containing three protons and two inner electrons between the protons, i.e., hydrogen-3.

Two-ending   A single deuteron that is placed on one side of a carbon nuclet.

Index